CAMBRIDGE LIBRARY COLLECTION

Books of enduring scholarly value

Physical Sciences

From ancient times, humans have tried to understand the workings of the world around them. The roots of modern physical science go back to the very earliest mechanical devices such as levers and rollers, the mixing of paints and dyes, and the importance of the heavenly bodies in early religious observance and navigation. The physical sciences as we know them today began to emerge as independent academic subjects during the early modern period, in the work of Newton and other 'natural philosophers', and numerous sub-disciplines developed during the centuries that followed. This part of the Cambridge Library Collection is devoted to landmark publications in this area which will be of interest to historians of science concerned with individual scientists, particular discoveries, and advances in scientific method, or with the establishment and development of scientific institutions around the world.

Memoirs of Dr. Joseph Priestley, to the Year 1795

Joseph Priestley (1733–1804) was an eighteenth-century English polymath with accomplishments in the fields of science, pedagogy, philosophy, and theology. Among his more notable achievements were the discovery of oxygen and his work in establishing Unitarianism. Often a controversialist, Priestley's efforts to develop a 'rational' Christianity and support for the French Revolution eventually made him unwelcome in his native land. His 1807 *Memoirs* relate the story of his life until the time of his 1794 emigration to America and include other biographical materials written by his son. This second volume contains a lengthy discussion of Priestley's theological writings and as well as four of his sermons. Priestley's memoirs are an important source for anyone interested in the state of epistemology, rationalism, and religious belief in the age of the Enlightenment, and in a man who, in the words of his son, 'gave unremitting exertions in the cause of truth'.

Cambridge University Press has long been a pioneer in the reissuing of out-of-print titles from its own backlist, producing digital reprints of books that are still sought after by scholars and students but could not be reprinted economically using traditional technology. The Cambridge Library Collection extends this activity to a wider range of books which are still of importance to researchers and professionals, either for the source material they contain, or as landmarks in the history of their academic discipline.

Drawing from the world-renowned collections in the Cambridge University Library, and guided by the advice of experts in each subject area, Cambridge University Press is using state-of-the-art scanning machines in its own Printing House to capture the content of each book selected for inclusion. The files are processed to give a consistently clear, crisp image, and the books finished to the high quality standard for which the Press is recognised around the world. The latest print-on-demand technology ensures that the books will remain available indefinitely, and that orders for single or multiple copies can quickly be supplied.

The Cambridge Library Collection will bring back to life books of enduring scholarly value (including out-of-copyright works originally issued by other publishers) across a wide range of disciplines in the humanities and social sciences and in science and technology.

Memoirs of Dr. Joseph Priestley, to the Year 1795

With a Continuation, to the Time of his Decease by his Son, Joseph Priestley

VOLUME 2

JOSEPH PRIESTLEY

CAMBRIDGE UNIVERSITY PRESS

Cambridge, New York, Melbourne, Madrid, Cape Town, Singapore,
São Paolo, Delhi, Dubai, Tokyo

Published in the United States of America by Cambridge University Press, New York

www.cambridge.org
Information on this title: www.cambridge.org/9781108014205

This edition first published 1807
This digitally printed version 2010

ISBN 978-1-108-01420-5 Paperback

MEMOIRS

OF

DR. JOSEPH PRIESTLEY,

TO THE YEAR 1795,

WRITTEN BY HIMSELF:

WITH A CONTINUATION, TO THE TIME OF HIS DECEASE,

BY HIS SON, JOSEPH PRIESTLEY:

AND OBSERVATIONS ON HIS WRITINGS,

BY THOMAS COOPER,

PRESIDENT JUDGE OF THE 4TH DISTRICT OF PENNSYLVANIA; AND THE REV. WILLIAM CHRISTIE.

TO WHICH ARE ADDED,

FOUR POSTHUMOUS DISCOURSES.

VOL. II.

LONDON:

PRINTED FOR J. JOHNSON, NO. 72, ST. PAUL'S CHURCH-YARD.

1807.

APPENDIX, N°. 6.

A Review of Dr. Priestley's Theological works, with occasional Extracts, expressive of his sentiments and opinions, and observations on his character and conduct as a Christian Minister.

"I can truly say, that I always considered the office of a Chris-
"tian Minister as the most honourable of any upon earth; and in the
"studies proper to it, I always took the greatest delight."

MEMOIRS, page 57.

WHILE some are usefully and commendably employed in celebrating the various merits and talents of a Priestley; in describing and discriminating with accuracy and skill the capacities and resources of his fertile and comprehensive mind, which, without perplexity or confusion, could embrace a variety of objects, and excel in experimental philosophy, metaphysics, philology, historical disquisitions, and speculations on civil government; be it my task (as far as my abilities can enable me to accomplish it) to trace and mark his progress as a Theologian, and to exhibit a brief, but faithful view,

 of

of those numerous productions that flowed from his pen, on subjects (as he justly thought) the most important and interesting of all others.

Intended and set apart, as he was, in the counsels of his nearest and best friends at an early period, for the Gospel Ministry, his own serious and devotional mind excited him to coincide with their views, and carried him forward with alacrity in the pursuit and attainment of his favourite employment, notwithstanding the embarrassments arising from a weak and delicate constitution, and the still greater difficulties that came in his way from the bigotry and hostility of those whose apprehensions of divine truth were different from his own.

Who can read the simple and artless narrative of his life, without admiration of the candour and ingenuity of the writer, who studiously lays open to the public view the circumstances of his birth and education, in which occurred some facts that the pride of many would have induced them carefully to conceal? Who can behold without indignation a Priestley struggling with poverty and contempt at Needham, and languishing on a salary of less than 30*l.* a year? What a just picture does he draw of the tem-

per-

per and disposition of too many persons in this present evil world, when he informs us, that when he came to preach at a certain place, the genteeler part of the audience carefully absented themselves; and that some time afterwards, when his character and fame had risen in the world, the very same persons came in crouds to hear him, and extolled a discourse that they had formerly slighted and despised!

The first Theological work he ever composed was his *Institutes of Natural and Revealed Religion*, the first part of which, he informs us in his Memoirs, he wrote at the academy: but as this work did not make its appearance till several years afterwards, I shall postpone my observations upon it till the period of its publication.

The first work he actually published was a treatise, entitled, *The Scripture Doctrine of Remission: which sheweth that the death of Christ is no proper sacrifice nor satisfaction for sin; but that pardon is dispensed solely on account of repentance, or a personal reformation of the sinner*, *London*, 1761. This piece was submitted to the perusal of Dr. Lardner, and Dr. Fleming, and by them published with the above title. The treatise originally took in a larger com-

pass,

pass, and comprehended *Remarks on the reasoning of the apostle Paul,* which he considered as inconclusive in some places. Dr. Lardner could not by any means approve of these remarks, and therefore they were left out in this publication, though afterwards committed to the press, and inserted in the Theological Repository. This treatise on *Remission* was composed at Needham, when the author could not exceed twenty-five years of age. It affords a singular proof of the strength of his genius, the clearness of his conceptions, the perspicuity of his style, and his familiar acquaintance with the language and phraseology of the sacred writers.

At the time of the Reformation, no rational theory at all existed with respect to the doctrine of remission, or the forgiveness of sin. The notions of the Popish schoolmen were implicitly adopted by the reformers, and their absurdity increased rather than diminished. The commentators on Scripture in general followed the same ideas. A degree of good sense indeed appears in Vatablus, as Socinus has justly remarked. The illustrious Faustus Socinus himself, before

before mentioned, was the first, properly speaking, that broke ground on the subject.

> Tota ruit Babylon; disjecit Tecta Lutherus,
> Calvinus Muros, et Fundamenta Socinus.

In his celebrated treatise *De Jesu Christo Servatore*, he has torn up the strong holds of school divinity and Calvinism, completely overturned the notion of a proper infinite satisfaction to the justice of God, and settled the true idea of Jesus Christ as a saviour, redeemer, mediator, and high priest, on a scriptural as well as a rational foundation. He was induced to compose this most elaborate and valuable performance, in consequence of a series of theological axioms and positions having been sent him by Covetus, a Calvinistic divine, who before had had a conference with him at Basil, in Switzerland, and wished to reclaim him from his supposed errors. Socinus wrote a large, distinct, and particular reply to all Covetus's objections, and forwarded it to him by the way of Geneva, where it fell into the hands of the Calvinistic divines of that place, who thought proper to detain it, apprehending it might have some bad effect upon the mind of Covetus himself, or others into whose hands it might have fallen. Fortunately however for the religious

religious world and himself, Socinus had kept a copy, which many years afterwards was published, with the author's improvements, and divided into four parts, by a friend of his in his own life-time; for though descended from a noble family, and heir to an estate in Tuscany, (by his attachment to the pure gospel of Jesus Christ) he was too poor to be able to publish it himself. The learned and eminent Hugo Grotius, many years after the death of Socinus, attempted to controvert what Socinus had advanced; but an excellent and judicious reply was made to him by Crellius, which had such an effect upon the honest and candid mind of Grotius, that he wrote Crellius a letter, giving him thanks, and acknowledging that he had set him right in some particulars. The subject of atonement was afterwards taken up by the Arminian divines of Holland, who attempted its defence on more moderate and tenable ground than that adopted by the Calvinists. Nearly on the same footing it was held by the semi-rational divines of the church of England, in the reign of Charles the Second and king William, viz. the Tillotsons, Burnets, and Stillingfleets. The last wrote a famous treatise at the time, (if I recollect right), entitled,

titled, *The true Reasons of the Sufferings of Christ*, which was animadverted upon occasionally by Mr. Emlyn, in his sensible and valuable reply to Lessley's dialogues against the Socinians. Mr. Biddle also discussed the subject of atonement, as did in a much later period Dr. Clarke, Mr. Tomkins, Hopton Haynes, Esq. Dr. Sykes, and Dr. Taylor, in his *Scripture Doctrine*, with many others. The subject also was well handled in an anonymous treatise, stiled, *The Scandal and Folly of the Cross removed; or, the wisdom of God's method of the Gospel, in the death of Jesus Christ, manifested and justified, against the Deists*, *London*, 1699.

It does not however appear at all probable to me, that young Mr. Priestley was acquainted at the time with the greatest part of the numerous writers before mentioned. I find no reference to any, except Dr. Clarke and Mr. Tomkins; Doctors Taylor and Sykes he might have seen, though this is not certain. Indeed, as he asserts in his Memoirs, and as he once assured me himself, he had recourse to the Scriptures at large, and carefully noted every passage in the old and new Testament, that he thought had any relation to the subject of his inquiry, and formed his judg-

ment

ment upon the whole collectively. The result was, that in his opinion there was *no atonement*. He has therefore, in some measure, all the merit of an original writer. In proof of the judgment he had formed, he urges many powerful arguments, supported by Scripture testimonies and striking considerations. I could with pleasure enter into a particular detail of his reasonings, if it were not that having afterwards inserted every sentence of this treatise in the Theological Repository, under the signature of *Clemens*, and also a considerable part in the *History of the Corruptions of Christianity*, under the head of *Atonement*, with many and valuable improvements, this detail will come in with more propriety when these works are spoken of. I shall content myself at present with extracting the Introduction, which contains some valuable observations, and a brief view of the schemes of different parties of christians respecting the doctrine of atonement, and affords an early specimen of the easy elegance of the author's stile; more especially as the Introduction, as far as I can trace, has never been republished, and is now very scarce.

"By reason of the poverty of all languages, the use of figurative expressions, or the affixing of the

same

same term to things that are only analogous to one another, cannot be avoided; especially, in treating of moral or religious subjects, in which our ideas themselves must necessarily, be much compounded, and borrowed from sensible things.

"What hath still more contributed to fill all languages with these artificial forms of speech, is, that when necessity had first introduced such an use of words, the ingenuity of men, as in other similar cases, presently worked it up into a beauty. Some allusions were observed to be so peculiarly happy and striking, as to incite men of taste and invention to seek for more: hence a language extremely scanty in its elements, comes to abound in words; most of which, however, are artificial or compounded, and may, with care, be reduced to their simple and component parts.

"But such is the nature both of our ideas and words, and such the power of association, that what was at first evidently compounded or figurative, by frequent use ceases to be conceived to be so: compound ideas and expressions in time pass for simple ones, till, after a rigorous scrutiny, their deviation be seen, and they appear to be factitious. In like manner, it is very possible to call one thing by the name of another by way of allusion only, till at last the allusion be forgotten, and the nature of the thing itself be mistaken.

"Though therefore, the derivation of words from so fruitful a source, does very much enrich a language, though the use of figures in speech, or writing, very much enliven a discourse, give a colour and strength to the expression, and, if the allusions be made with judgment, may, in many cases, facilitate the discovery of truth; yet the too free an use of them may embarrass the sense, and render the speaker or writer's meaning very dubious or obscure, especially to those who are not used to his manners.

"This is universally complained of where the writings of the *Asiatics* in general fall into the hands of *Europeans;* they go so far beyond us in the boldness and freedom of their figures. And this is one cloud that hangs over the true meaning of the writers of the books of scripture; which, at this distance of time and difference of manners, it is exceeding difficult for us to see through, and hath led their readers into very widely different apprehensions of their sense, some resting in the most obvious and gross meaning of of the words they use; while others, suspecting this to be falling short of their true meaning, wander many different ways in quest of it.

" Perhaps, concerning no one thing of which the sacred writers do treat, have the notions of moderns been more widely different, than concerning the account they give us of the death of Christ; the view they supposed he suffered with, and the end, which they assert, was in part answered by it. The most distinguished opinions that are maintained among christians at this day, seem to be the following.

" *First*, some maintain that Christ, in his agony and death, endured pains equal in degree (the dignity of his person considered) to those that sinful men ought to have suffered on account of their sins, by a kind of substitution of persons, and transferring of guilt; agreeable to which, they hold, that this was the proper notion of a sacrifice for sin under the law; all which, they say, were *types* or *emblems* of the great sacrifice of Christ. But of those who agree with them that the pardon of sin is dispensed in consideration of the sufferings of Christ, all do not insist that the Divine Being could not, consistently with the honour of his perfections, have accepted of less

than

than a full equivalent for satisfaction; some supposing the Divine Being to have been at liberty to accept of any finite satisfaction that he pleased.

"*Secondly*, others again, agree with the former, that the death of Christ is a proper sacrifice for sin, like the Jewish sacrifices, but then they suppose, that the virtue of a sacrifice consisted, not in the shedding of the blood, or the death of the victim, but in the disposition of the offerer, of which the sacrificing of the beast was in some manner emblematical; and that in like manner the virtue of the death of Christ consisted, not in the pain that he endured, but in his real virtue and worth, manifested to God and the world by his obedience unto death. Though, therefore, they deny the necessity of any vicarious sufferings, they assert the necessity of the interposition and mediation of some person of distinguished virtue and worth, on the part of the offenders, before the Divine Being could in wisdom, dispense pardon to them.

Lastly, others, in direct opposition to both the before mentioned opinions, maintain, that the death of Christ had no manner of relation to a proper sacrifice for sin; and that the apostle never meant more than a figurative allusion to those Jewish rites: that Christ

Christ died in consequence of his undertaking to reform a vicious world, for the proof of his divine mission and doctrine, and other such rational, simple, and consistent ends. They maintain that there was no necessity for satisfaction of any kind, or the interposition of any being whatever, in order to God's remitting the sins of men."

An interval of six years took place after the publication of the before mentioned treatise on Remission, in 1761, before Dr. Priestley appeared again from the press in the character of a Theologian. His time, as he says, had been occupied with the business of teaching at Nantwich and Warrington. But in 1767, when he was again settled as a minister at Leeds, he resumed his theological studies with fresh ardour. The effect of this application appeared in various publications, which followed one another almost in constant succession; and while they rendered his name celebrated in the religious world, drew on him a storm of obloquy and reproach. About the same time, in 1767, came abroad his *Catechisms* for children and young persons, *Scripture Catechism*, *Forms of Family Prayer*, and Treatise on the *Lord's Supper*. The three first of these are plain useful

pieces,

pieces, exceedingly well calculated to promote the instruction and improvement of youth in principles of piety and virtue, and to excite and enable masters of families to the performance of the too much neglected, but highly necessary duty of family prayer. The Catechisms are remarkable for their simplicity and freedom from all points of controversy, and on this account may be safely used by christians of very opposite sentiments. A knowledge of the most important facts in holy writ may be acquired in early youth by the careful use of the Scripture Catechism. The last mentioned piece, entitled, *A Free Address to Protestant Dissenters, on the subject of the Lord's supper*, will deserve a more particular consideration.

The superstition of the Papists, and the absurdities attending the doctrine of transubstantiation, are sufficiently known to all Protestants, and justly and universally condemned. But have Protestants themselves kept clear of all false ideas and improprieties in their conceptions and administration of this institution? Luther held a half-way or compromising doctrine on this subject, called *consubstantiation*. Calvin avoided this error, but inculcated notwithstanding, notions that will not bear the test of reason, scripture,

or

or experience. The English reformers, Cranmer and others, adopted similar superstitious and unwarrantable ideas. Zuinglius, and a few others, appear, from the account given of their tenets, to have come pretty near the truth; and Faustus Socinus, with that penetration and sagacity which generally accompanied him in theological disquisitions, has in his tract *De usu et fine cænæ domini*, in a great measure explained this institution, according to the simplicity in which it is found in scripture. The other great men who succeeded him in the Unitarian churches of Poland and Transylvania, have followed the same method of interpretation with little variation. But these bright luminaries for a long time shone in vain. They were unable to dispel the general darkness in which the christian world was enveloped. Of the many tracts published by churchmen and dissenters, before the year 1730, none keep clear of extremes on this point. The best of them that I have seen is that published by the pious and worthy Mr. Henry Grove. It was reserved for bishop Hoadley to throw full light upon this subject, and exhibit it in all its scriptural simplicity; though he did not escape censure for so doing, but was severely animadverted upon by the Waterlands and William Laws of those days.

Dr. Priestley following the plan of bishop Hoadley; and exerting his own good sense at the same time, composed an excellent and edifying treatise on the Lord's supper, to which a very sensible preface is prefixed, exhorting and animating Protestant dissenters to a free and impartial examination of this and other religious topics, to consider the importance and advantages of their situation, and make a suitable improvement of them. In treating the subject, he first recites the accounts the three first evangelists have delivered of the institution, and also that of the apostle Paul. He insists more particularly upon this last, and shews that the kind of unworthy communicating which Paul censures, and warns against, does not relate to any failure in those preparations which so many lay an undue stress upon, but in eating and drinking to excess on the occasion; and not distinguishing between the Lord's supper, and an ordinary meal or common feast. He then adds, " All the censure that St. Paul passes upon unworthy communicants, I would observe by the way, relates wholly to such a manner of receiving this ordinance, as is no where practised at this day in any christian country. His censures, therefore, are evidently such as no christians at this day can justly apply to themselves." He defines the Lord's supper to be, "a solemn, but chearful rite, in remembrance of Christ, and of what he has done and suffered for the benefit of mankind. Like other customs, which stand as records of past events, it preserves the memory of the most important of all transactions to the end of the world, even till Christ's second coming." He proceeds, "If I be asked, what is the *advantage* of celebrating this rite; I answer, it is of the same nature as that which results from repeating any custom, in commemoration of any other important event; of the same nature with the celebration of the passover, for instance, among the Jews. It tends to perpetuate the memory of the transaction recorded by it, and to cherish a grateful and joyful sense of it. In this case, the custom tends to perpetuate the memory of the death of Christ, and to cherish our veneration and love for him. It inflames our gratitude to so great a benefactor, and consequently our zeal to fulfil all his commands.

"The celebration of the Lord's supper being, more especially, a commemoration of his *death*, it serves to remind us that we are the disciples of a crucified master, and it is therefore a means of fortifying our minds, and preparing them for every degree of hardship and persecution to which we may be exposed in the profession of christianity. It reminds us that we are *not of this world*, even as our Lord was not of it, and (*servants not being greater than their Lord*) that we have no right to expect better treatment from the world, than he met with from it. By this means it serves to keep up in our minds a constant view to the great object and end of our christian profession, viz. *the expectation of a future life*, and to cherish the mortification to the world, and that heavenly mindedness, which are eminently useful in fitting us for it.

"On these occasions then, more especially, let us reflect, that if, in the hour of temptation, we deny Christ, *he will also deny us*; that if in circumstances of reproach, we be ashamed either of the profession of his gospel, or of that strictness and propriety of conduct to which it obliges us, *he also will be ashamed of us* in that great day when he shall come *in his own glory*,

glory, in that of his father, and of his holy angels; but that if, we steadily and uniformly *confess him before men*, by an unblameable life and conversation, and by proper fortitude in bearing the trials to which we may be exposed for the sake of his truth, and of a good conscience, *he also will confess us before his heavenly father and the holy angels;* so that *if we suffer with him, we shall also reign with him, and be glorified together.*

"Lastly, the celebration of the Lord's supper being the joint action of several persons, it strengthens our affection to the common cause, to one another, and to all who are engaged in it. If you expect more than this, your expectations are unreasonable, enthusiastical, and sure to be disappointed." The rest of this section, and the next, more particularly treats of the qualifications of communicants, and cautions against excessive rigour in judging of the fitness of persons for partaking of this ordinance. Section third contains observations on the Lord's supper, being called a sign, or seal of the covenant of grace. The fourth section exhibits a brief history of the corruptions of the christian doctrine and practice with respect to it among the early Christians, the Reformers,

the

the English and Scotch establishments, and the Protestant dissenters. The fifth and last section contains an earnest and animated exhortation to all christians to the celebration of this institution, in a devout, serious, and rational manner, as a grateful and affectionate memorial of their great and generous benefactor Jesus Christ.

No man can labour with energy and effect in the cause of reformation, but he will more or less excite the resentment of those who either cannot, or will not enter into his views. Dr. Priestley's well intended attempt to enlighten the mind of dissenters with respect to the Lord's supper, drew upon him a rude and illiberal attack from Mr. Venn, a clergyman of the church of England, to which he replied with such calmness, moderation, and delicate irony, that his reply may be held up as a model of christian temper and fortitude, in return for harsh censure and ill usage. It bears the title of *Considerations on differences of opinion among Christians, with a letter to the Rev. Mr. Venn, in answer to his free and full examination of the Address to Protestant Dissenters, on the subject of the Lord's supper*, London, 1769.

I think it needless to enlarge upon the letter in which

which Mr. Venn is properly confuted; but these considerations are so replete with valuable matter, that they deserve to be attended to. They were again reprinted at Birmingham in 1790, and subjoined to *Familar Letters addressed to the Inhabitants of Birmingham*, &c." These considerations are divided into five sections. The first treats of latent insincerity and direct prevarication, and points out the sources of deception by which controversial writers and leaders of sects and parties impose upon themselves, and come under the influence of motives that they are scarcely conscious of. The second inquires into the source of bigotry and persecution, which arises chiefly from a blind and violent attachment to particular opinions, and connecting the only possibility of salvation with the belief of them. The third describes the practical tendency of different systems of doctrine, in which it is shewn that the great objects of hope and fear, which christianity presents to mankind, viz. the joys and torments of a *future life*, must be nearly the same in all the forms of the christian religion, and in proportion to the degree in which we give our attention to them, and thereby strengthen our faith in them, they must

influence

influence us all alike. All the difference, therefore, with respect to the practical influence of any particular opinions, can only be occasioned by the different views with which they present us, of those *persons* and *things* that are objects of our duty. A brief illustration of this thought is afterwards given in the idea that is exhibited of the Divine Being, according to what are generally called the *rational* and the *orthodox* systems. The comparison, which as far as I am able to judge, is a fair one, turns out by no means in favour of the orthodox system, the tendency of which appears to be to something else than virtue: though as the author candidly remarks, better principles (which he states) *really*, though secretly influence the conduct of those who are truly pious and virtuous among Calvinists; and by no means the principles which they profess.

The fourth section assigns the causes of difference of opinion, and recommends the reading of the scriptures. What our author says here appears to me of such prime importance, and so much for the interest of all christian sects to attend to, that I think myself bound to quote the whole of it.

"I cannot

"I cannot help wishing that persons of all sects and parties would study their bibles *more*, and *books of controversy* less. But all persons have their *favourite authors*, to which they too much confine themselves, even to the neglect of those *authorities*, from their agreement with which, all their merit is acknowledged to be derived. Were it not for this circumstance, it would be absolutely impossible that the individuals of mankind, whose intellects are so much alike, should differ so widely in their religious sentiments as they now do; at least that they should lay so great a stress on the points in which they differ.

"Since the understandings of men are similar to one another, (at least so much so, that no person can seriously maintain, that *two* and *two* make *five*) did they actually read only *the same books*, and had they no previous knowledge to mislead them, they could not but draw the same general conclusions from the same expressions. But one man having formed an hypothesis from reading the scriptures, another, who follows him, studies that hypothesis, and refines upon it, and another again refines upon him; till, in time, the scriptures themselves are little read by any

of

of them, and are never looked into but with minds prepossessed with the notions of others concerning them. At the same time, several other *original readers* and thinkers, having formed as many other hypotheses, each of them a little different from all the rest, and all of them being improved upon by a succession of partisans, each of whom contributed to widen the difference; at last no religions whatever, the most distinct originally, are more different from one another, than the various forms of *one* and the same religion.

"To remedy this inconvenience, we must go back to first principles. We must begin again, each of us carefully studying the scriptures for ourselves, without the help of commentators, comparing one part with another. And when our minds shall, by this means, have been exposed for a sufficient time, to the same influences, we shall come to think and feel in the same manner. At least, all christians being sensible that they have, in many, and in the chief respects, *one Lord, one faith, one baptism*, they will keep *the unity of the spirit, in the bond of peace.*

"In

"In reading books of controversy, the particular texts from which favourite opinions are chiefly inferred, are kept continually in view, while others are kept out of sight; so that the person who confines himself to the perusal of them, necessarily forms a very strong notion of the *general tenor* of the scriptures, and lays a disproportionate stress on particular opinions. He never looks into the scriptures, but it is with a state of mind that leads him to expect to find his opinions either clearly expressed, or plainly referred to in every chapter. Now, it is well known, that all strong expectations tend to satisfy themselves. Men easily persuade themselves that they actually see, what they have absolutely depended upon seeing.

"Were it possible for a number of persons to make but an essay towards complying with this advice, by confining themselves for the compass of a single year, to the daily reading of the scriptures only, without any other religious books whatever; I am persuaded that, notwithstanding their previous differences, they would think much better of one another than they had done before. They would all have, more nearly, the same general ideas of the contents of scripture, and of the chief articles of christian faith and duty. By reading the whole themselves, they could hardly avoid receiving the deepest impressions of the certainty and importance of the great and *leading principles*, those which they would find the most frequently and earnestly inculcated; and their particular opinions having come less frequently in view, would be less obstinately retained. It was in this manner, I can truly say, that I formed the most distinguishing of my opinions in religion.

"I do not say that this practice would have the same effect with all persons. I have no hopes of its succeeding with those who are advanced in life. I would not even recommend it to them; since the consequence of unhinging their minds, though by a conversion from error to truth, might possibly do them more harm than good. Nor have I much hope of those who are hackneyed in controversy, and to whom the methods of attack and defence, peculiar to any system, are become familiar. But I would earnestly recommend this method of studying the scriptures to young persons, before their common sense and natural feelings have been perverted; and

while

while they are capable of understanding the obvious meaning of a plain expression.

"In this case I cannot help thinking, that notwithstanding the seeming force of the texts that are continually in the mouths of those who call themselves *orthodox*; and notwithstanding our present translation of the bible, which (being made by men who were fully persuaded of the truth of that system) is, in many places, much too favourable to it; yet that both the general *tenor of the whole*, (which, with a person who reads the scriptures much, cannot but have far greater weight than any particular texts whatever) and also that number of *emphatical single passages*, would effectually over-rule any tendency to that which is commonly called orthodoxy.

"To mention a single instance. Would not a constant attention to the general strain in which Moses, all the antient prophets, John the Baptist, our Saviour, and the apostles, wear out, in time, every trace of the doctrine of unconditional election and reprobation? The language in which the Divine Being is uniformly represented in speaking is, *As I live, saith the Lord, I would not the death of a sinner, but had rather that he would repent and live. Turn ye, turn*

ye,

ye, why will ye die, oh house of Israel. What a solemn and cruel mocking of mankind would this be, if the Divine Being, at the same time that he made this declaration, was purposed that many, if not the greatest part of them, should not repent, but die without mercy?"

The fifth and last section contains *general advice*, which deserves to be carefully read and reduced to practice.

Mr. Venn was not the only opponent Dr. Priestley had to encounter in consequence of his treatise on the Lord's supper. Nine letters were published by an anonymous author, under the title of *The Protestant Dissenter's Answer to the Free Address on the subject of the Lord's supper*. In a letter addressed to the author, Dr. Priestley replies to him, in which he makes the following candid acknowledgment. "I thank you because you have led me, as you will see, to correct some mistakes, and to amend some expressions which had inadvertently escaped me, and more especially to make such *additions* to what I had written as appear to me to be favourable to my original and professed design in writing." Notwithstanding this concession, he considers himself as in the

the right with respect to the general plan and execution of the work, and defends what he had advanced on the subject. Nor does it appear that the author of these nine letters differed materially from him in his notion of the Lord's supper. In the view of the author entering into a farther discussion of the subject, he states seven queries, and proposes them to his consideration.

That I may bring together under one point of view all that Dr. Priestley has written on the subject of the Lord's supper, I shall here give an account of a tract, though a little out of the order of time, entitled, *An Address to Protestant Dissenters, on the subject of giving the Lord's supper to children*, London, 1773.

Our author informs his readers, that having been more conversant with the antient christian writers called *Fathers*, and especially having met with Mr. *Peirce's Essay* on the subject, he is now, upon mature consideration, fully satisfied, that *infant communion*, as well as *infant baptism*, was the most antient custom in the christian church; and therefore that the practice is of apostolical, and consequently of divine authority. His chief arguments in in favour of infant communion are, that infants were capable of full communion under the Jewish dispensation, having been not only circumcised, but partaking also of the passover; and that infant communion, as well as infant baptism, was the practice of the primitive christians. He proves this last assertion from the testimonies of Cyprian and Austin, and manages and illustrates his arguments with great dexterity. He shews that infant communion continued for a long time in the church of Rome, and was not forbidden by an express and formal determination of any council, till the fifteenth century, at the council of Basil, some time after they had, in the council of Constance, in 1415, decreed, that the laity should receive the communion in one kind only. But that infant communion is to this day the practice of the Greek churches, of the Russians, the Armenians, the Maronites, the Copts, the Assyrians, and probably all other oriental churches; and it was also the practice of the Bohemians, who kept themselves free from Papal authority till very near the reformation. In conclusion, he points out the advantages which might arise from returning to the use of this antient custom. But previous to this he observes,

serves, that since the administration of the Lord's supper is an act of public worship, the ends of the institution will be answered, if children be brought to communion as soon as it is found convenient for them to attend other parts of public worship. First, were children admitted to the Lord's supper, they would become more the objects of attention, both to their parents and the governors of churches, and greater care would be taken of their christian education. They themselves also would be more apt to inquire concerning the reasons of what themselves constantly did, and thus furnish an easier handle for their religious instruction.

Secondly; the principal advantages that might be expected from it is, that, by this means, young persons would probably be more firmly established in the belief of christianity. Having been from their infancy, constantly accustomed to bear their part in all the rites of it, they would be more firmly attached to it, and less easily desert it, &c.

Thirdly, the revival of the practice of infant communion might be a means of reviving an useful *church discipline*, which is altogether lost among us,

and

and of the want of which many wise and good men complain, &c.

Let not any man pass a premature censure upon Dr. Priestley's judgment in this particular, from the abridged view of his arguments here exhibited, without having recourse to the pamphlet itself, which contains much farther information on the subject.

We are now to contemplate Dr. Priestley under a new character, as the intrepid champion of the Protestant dissenters, standing forth in vindication of their just rights and privileges, against the exorbitant claims of high churchmen and the imperious usurpation of interested priests, laying before them the importance and advantages of their situation, proposing to their imitation the example of their heroic ancestors, and animating them to a conduct and behaviour, in all respects worthy of real christians, and enlightened and conscientious dissenters.

A long controversial war had existed, and been carried on with much clamour and obloquy between the advocates of diocesan episcopacy and the Puritans, Presbyterians, and other classes of the dissenters in England, almost from the reformation down to the accession of the present reigning family. Innumerable

numerable books and pamphlets had been written concerning the authority of the church, the power of the clergy, the apostolical succession, the *jure divino* right of episcopacy, &c. The Puritans and Dissenters were not wanting on their part in producing a number of replies, in some of which the *jure divino* right of presbytery, or other forms of church government, were maintained in opposition to the Episcopalians.

Soon after the accession before mentioned, the controversy began to take a different turn. The influence of philosophy, the love of religious liberty, the spread of the maxims of toleration, and above all the diffusion of rational theology, brought ecclesiastical jargon into contempt, and subdued the ferocity of fiery polemics. The priests considered as a body were either confuted or laughed out of their absurdities. Add to this, that the first princes of the house of Brunswick, and their state counsellors, were themselves low churchmen, and from political and other motives disposed to patronize moderate men and moderate measures, and favourably inclined to dissenters.

In this state of things some of the more intelligent of the clergy, sharing no doubt in the general illumination, and finding that the old priestly dogmas would not now serve their turn, or go down smoothly with the improved part of the nation, saw the necessity of framing a new hypothesis on which to raise the precious fabric of clerical domination, and give a new currency to the *wares* of Babylon. The acute and subtile genius of a *Warburton*, was deemed adequate to the task. His *alliance between church and state*, came forth like a stalking horse to attract the admiration of the croud, to dazzle weak minds, and make *the worse appear the better reason.* It was to be expected that men of inferior abilities would copy from so great a master, and that various modifications of the general principles of this work would be attempted. Dr. Balguy, in a sermon he published on the subject of church authority, asserted, that it greatly concerned the public peace and safety, " that all church authority should be under the " controul of the civil governor; that religious as" semblies as well as others, should be subject to his " inspection, and bound by such rules as he should see " fit to impose." And that " the most effectual " method

"method of obtaining this security, was to invest the "supreme power, civil and ecclesiastical, in the same "person." He maintains in the same discourse, the obligation of the civil magistrate to establish the religion of the majority of his subjects, even though he might not be convinced that it was the best form of religion. Against these positions, and others connected with them or flowing from them in the judgment of this writer, Dr. Priestley thought it became him to publish a reply, bearing the title of *Considerations on Church Authority, occasioned by Dr. Balguy's Sermon on that subject; preached at Lambeth chapel, and published by order of the Archbishop*, London, 1769.

The work is divided into six sections. In the four first he embraces a larger scope than that suggested by Dr. Balguy's discourse, and argues against the different forms of priestcraft and church authority in general, confuting with masterly skill the sophistry and subterfuges that have been used in their defence. In the two last sections he confines himself chiefly to Dr. Balguy's positions and manner of reasoning, which he refutes in a solid manner. In this work there are *verba ardentia*, glowing forms of expression,

expression, and ingenious arguments, which would well deserve to be held up to public view, and would adorn these pages very much; but my limited plan will only permit me to extract a few of them, and these will in some degree suffer by being separated from what goes before and what follows after.

Page 4. "All the civil societies we enter into in this life will be dissolved by death. When this life is over, I shall not be able to claim any of the privileges of an Englishman; I shall not be bound by any of the laws of England, nor shall I owe any allegiance to its sovereign. When, therefore, my situation in a future life shall have no connection with my privileges or obligations as an Englishman, why should those persons who make laws for Englishmen interfere with my conduct, with respect to a state to which their power does not extend?"

P. 5. "As a being capable of immortal life, (which is a thing of infinitely more consequence to me than all the political considerations of this world) I must endeavour to render myself acceptable to God, by such dispositions and such conduct as he has required, in order to fit me for future happiness. For this purpose, it is evidently requisite, that I diligently

gently use my reason, in order to make myself acquainted with the will of God; and also that I have liberty to do whatever I believe he requires, provided I do not molest my fellow creatures by such assumed liberty. But all human establishments, as such, obstruct freedom of inquiry in matters of religion, by laying an *undue bias* upon the mind, if they be not such, as by their express constitution prevent all inquiry, and preclude every possible effect of it.

"Christianity, by being a more spiritual and moral constitution than any other form of religion that ever appeared in the world, requires men to think and act for themselves more accurately than any other. But human establishments, by calling off men's attention from the commandments of God to those of men, tend to defeat the great ends of religion. They are, therefore, incompatible with the genius of christianity."

P. 10. "By *the gospel*, every christian will, and must understand, the gospel in its purity; that is, what he apprehends to be the purity of the gospel, in opposition not only to heathenism, and religions *fundamentally false*, but to erroneous christianity, or to religions that are *in part true*. Whatever be the religious

ligious opinions, therefore, that I seriously think are agreeable to the word of God, and of importance to the happiness of mankind, I look upon myself as obliged to take every prudent method of propagating them, both by the use of speech and writing; and the man who refrains from doing this, when he is convinced that he should do good upon the whole by attempting it, whatever risque he might run in consequence of opposing anti-christian establishments, is a traitor to his proper lord and master, and shews that he fears more *them who can only kill the body*, (whether by the heathen methods of beheading, crucifying, throwing to the wild beasts, &c. or the christian methods of burning alive, and roasting before a slow fire) than him, *who can cast both soul and body into hell.*

"It is said by some, who think themselves obliged to vindicate the conduct of Christ and his apostles, that, though no general plea to oppose an established religion can be admitted, in excuse of a pretended reformer, yet that a *special* plea, such as a belief of a divine commission, and the like, will excuse him. But I can see no material difference in these cases. The voice of *conscience* is, in all cases, as the *voice of*

God

God to every man. It is, therefore, my duty to enlighten the minds of my friends, my countrymen, and mankind in general, as far as I have ability and opportunity; and to exert myself with more or less zeal in proportion as I myself shall judge the importance of the occasion requires, let my honest endeavours be considered as ever so factious and seditious, by those who are aggrieved by them. It is no new cry among the enemies of reformation, *the men who have turned the world upside down, are come hither also.* There are some who confine the obligation to propagate christianity to the *clergy*, and even to those of them who have *a regular commission* for that purpose, according to the form of established churches; and say that *laymen* cannot be under any obligation to trouble themselves about it, in whatever part of the world they be cast; and what they say concerning the propagation of christianity, they would extend to the reformation of it. But I can see no foundation for this distinction, either in reason or in the scriptures. The propagation, or reformation of christianity, is comprehended in the general idea of *promoting useful knowledge* of any kind, and this is certainly

certainly the duty of every man in proportion to his ability and opportunity.

"Our Saviour gives no hint of any difference between *clergy* and *laity* among his disciples. The twelve apostles were only distinguished by him as professed witnesses of his life, death, and resurrection. After the descent of the Holy Ghost, supernatural gifts were equally communicated to all christian converts. The distinction of *elders* was only such as years and experience entitled men to, and only respected the internal government of particular churches. As to the propagation of christianity abroad, or the reformation of corruptions in it at home, there is nothing in the scriptures that can lead us to imagine it to be the duty of one man more than another. Every man who understands the christian religion, I consider as having the same commission to teach it that I myself have; and I think my own commission as good as that of any bishop in England, or in Rome."

P. 18. "It is allowed by many, that christian churches as such, and its offices as such, have no right to inflict civil punishments; but they say the civil magistrate may embrace the christian religion,

and

and enforce its precepts by civil penalties. But have civil magistrates, when they become christians, a power of altering or new modeling the christian religion, any more than other members of the christian church? If not, its laws and sanctions remain just as they did before, such as Jesus Christ and his apostles left them; and the things that may have been substituted in their place cannot be called christianity, but are something else.

"If the civil magistrate chuse to become a christian, by all means let the doors of the christian church be open to him, as they ought to be to all, without distinction or respect of persons; but when he is in, let him be considered as no more than any other private christian. Give him a vote in all cases in which the whole assembly is concerned, but let him, like others, be subject to church censures, and even to be excommunicated or excluded for notorious ill behaviour.

"It is, certainly, contrary to all ideas of common sense, to suppose that civil magistrates embracing christianity have, therefore, a power of making laws for the christian church, and enforcing the observance of them by sanctions altogether unsuitable to its nature. The idea cannot be admitted without supposing a total change in the very first principles and essentials of christianity. If civil penalties be introduced into the christian church, it is, in every sense, and to every purpose, making it *a kingdom of this world.* Its governors then assume a power over men's persons and property, a power unknown in the institutes of our religion. If, moreover, the civil magistrate take upon him to prescribe creeds and confessions of faith, as is the case in England, what is it but to usurp a *dominion over the faith of christians?* a power which the apostles themselves expressly disclaimed."

P. 33. "Had there been such a connection between ecclesiastical and civil matters, as the advocates for church power contend for; had it been the proper office of the civil magistrate to superintend the affairs of religion, and had it been unlawful, as some assert, for private persons to attempt any alteration in it, except by application to the civil governor, is it not unaccountable, that our Lord, and his apostles, did not make their first proposals to the supreme magistrates among the Jews or Romans? They certainly had no idea of the peculiar obligation of magistrates

trates to attend to this business, and chuse a religion for the people, since we never hear of their making application to them on any such account. It was their constant custom to preach the gospel wherever they came, in all companies, and to all persons promiscuously; and almost all the intercourse they had with magistrates, seems to have been on occasion of their being brought before them as criminals.

"Our Lord sent out, both his twelve apostles, and also the seventy disciples, among all the cities of Israel, but we do not read of his sending any deputation to the rulers of the Jews. John the Baptist seems to have confined his preaching to the wilderness of Judea, and the territory in the neighbourhood of the river Jordan; where he gave his exhortation to all that came to hear him without distinction of persons. St. Paul, indeed, made an appeal to Cæsar, but it was in order to obtain his liberty in an unjust prosecution. We are not informed that he, or any of the apostles, ever took any measures to lay the evidences of the christian religion before the Roman emperor, or the Roman senate, in order to convince them of the truth and excellency of it, and induce them to abolish heathenism, in favour of it,

throughout

throughout the Roman empire; which many persons would now think to have been the readiest, the most proper, and the best method of christianizing the world. On the contrary, their whole conduct shews, that they considered religion as the proper and immediate concern of every single person, and that there was no occasion whatever to consult, or advise with any earthly superior in a case of this nature."

P. 35. "It cannot be inferred from any thing that our Saviour has delivered, that any one christian has a right authoritatively to dictate or prescribe to another, but I think the very contrary, if it be in the power of words to convey such a meaning. When his disciples were disputing about power and precedency, he said to them, Matth. xxiii. 8. *Be not ye called Rabbi, for one is your master, even Christ, and all ye are brethren; and call no man your father upon earth, for one is your father, which is in heaven. Neither be ye called master, for one is your master, even Christ; but he that is greatest among you shall be your servant*, &c. Mark x. 42. *Ye know that they who are accounted to rule over the Gentiles, exercise lordship over them; and their great ones exercise authority upon them; but so shall it not be among you;*

but

but whosoever will be great among you, shall be your minister, and whosoever of you will be chiefest, shall be servant of all."

P. 43. "All the rational plea for ecclesiastical establishments, is founded on the necessity of them, in order to enforce obedience to civil laws; but though religious considerations be allowed to be an excellent aid to civil sanctions, it will not, therefore, follow, as some would gladly have it understood; that, therefore, the business of civil government could not have been carried on *at all* without them. I do not know how it is, that this position seems, in general, to have passed without dispute or examination; but, for my own part, I see no reason to think that civil society could not have subsisted, and even have subsisted very well, without the aid of any foreign sanctions. I am even satisfied that, in many countries, the junction of civil and ecclesiastical powers have done much mischief, and that it would have been a great blessing to the bulk of the people, if their magistrates had never interfered in matters of religion at all, but had left them to provide for themselves in that respect, as they do with regard to medicine."

The

The state of things in this country since the American revolution, has justified the observations of our author here, and in other places. Civil government is found to subsist very well, and to answer all the purposes of society in Virginia, Pennsylvania, and in general throughout the United States, without the assistance of an incorporated band of clergymen and the sanction of a religious establishment.

P. 49. "Though it may be true, that inconvenience would arise from the immediate suppression of religious establishments, it doth not follow that they were either necessary or expedient; that the nation would have been in a worse state if they had never existed; and that no measures ought to be taken to relax or dissolve them. Were the religion of Mahomet abolished, every where at once, no doubt much confusion would be occasioned. Yet what christian would, for that reason, wish for the perpetuity of that superstition? The same may be said of Popery, and many other kinds of corrupt religion. Customs, of whatever kind, that have prevailed so long as to have infuenced the genius and manners of a whole nation, cannot be changed without

out trouble. Such a shock to men's prejudices would necessarily give them pain, and unhinge them for a time. It is the same with vicious habits of the body, which terminate in diseases and death; but must they be indulged, and the fatal consequences calmly expected, because the patient would find it painful and difficult to alter his manner of living? Ecclesiastical establishments, therefore, may be a real evil, and a disease in civil society, and a dangerous one too, notwithstanding the arguments for the support of them, derived from the confusion and inconvenience attending their dissolution; so far is this consideration from proving them to be things excellent or useful in themselves.

" Even the mischiefs that might be apprehended from attempts to amend or dissolve establishments, are much aggravated by writers. Much less opposition, I am persuaded, would arise from the source of real *bigotry*, than from the quarter of *interest*, and the bigotry that was set in motion by persons who were not themselves bigots."

P. 52. " One circumstance in favour of my argument is very evident. If the support of christianity had not been piously undertaken by Constantine,

tine, and the succeeding Roman emperors, the Popish hierarchy, that great *mystery of iniquity* and *abomination*, could never have existed. And I think all the advocates for church power, will not be able to mention any evil attending the want of ecclesiastical establishments, equal to this which flowed from one.

"All other ecclesiastical establishments among christians, partake more or less of the nature of this, the first and greatest of them being nothing more than corrections and emendations of it. Many of the abuses in it have been rectified, but many of them, also, are retained in them all. That there are some things good and useful in them all is true, but it is no difficult matter to point out many things that are good (that is, which have been attended with consequences beneficial to mankind) in the grossest abuses of popery. Those who study history cannot fail to be acquainted with them, and there is no occasion to point them out in this place.

"Thanks to the excellent constitution of things, that there is no acknowledged evil in the whole course of nature, or providence, that is without a beneficial operation, sufficient to justify the appointment or permission

permission of it, by that great and good Being who made, and superintends all things. But because tempests by land and sea, poisonous plants and animals, &c. do good, considered as parts of the whole system; and because it certainly seems better in the sight of God, that they should exist than not, must we not, therefore, guard against their pernicious effects to ourselves?

"Let this be applied to the case of civil and ecclesiastical tyranny in every form. The Divine Being, for good and wise ends, permits them; but he has given us a power to oppose them, and to guard ourselves against them. And we need not doubt, but that things will be so guided by his unseen hand, that the good they were intended to answer will be answered, notwithstanding our just opposition; or will appear to *have been* answered, if we succeed in putting a final end to them. He makes use of men as his instruments, both in establishing and removing all these abuses, in civil and ecclesiastical government."

P. 69. "I am afraid our Saviour and his apostles were not aware of this necessity of a legal maintenance for gospel ministers, or they would certainly ly have made some provision for it, or have left some instructions concerning it. But, perhaps, this was omitted by them, to prevent any reflection being cast upon themselves; for, according to this principle, they were but indifferently qualified for the discharge of their office. To be perfectly serious: If our Lord had imagined that any real advantage would have accrued to the ministers of his gospel from a legal provision, I do not see why we might not (either in his discourses or parables) have expected some hint of it, and some recommendation of an alliance of his kingdom with those of this world, in order to secure it to them. But no idea of such policy as this can be collected from the New Testament. For my part, I wonder how any man can read it, and retain the idea of any such worldly policy, so far am I from thinking it could have been collected from it."

In the same year, 1769, Dr. Priestley found a new and eminent antagonist against whom to exercise his talents, in defence of the rights of Protestant dissenters. Dr. Blackstone, the celebrated author of the Commentaries on the laws of England, had not only recited with approbation the statutes of Edward VI. and

and Queen Elizabeth, in which the penalty of confiscation of goods and imprisonment for life, for the third offence, are denounced against all who shall speak in derogation, &c. of the book of common prayer, but justified the continuance of such penalties, intimating that any alteration of them would be a breach of the articles of union between England and Scotland, and censuring in harsh and severe language, every attempt to depreciate the liturgy, as calculated for no other purpose than merely to disturb the consciences, and poison the minds of the people.

Dr. Priestley, in a bold and manly reply, and with a more than ordinary vehemence, which he thought the occasion called for, as believing himself to be particularly aimed at, refutes what Dr. Blackstone had advanced, and points out the injustice of such statutes, and the illiberality of those who undertake to defend them; inasmuch as dissenters are thereby precluded from making a proper defence of their principles, which can never be done with energy or effect, without exhibiting the true grounds of their dissent, founded on the unscriptural forms of worship contained in the books of common prayer. He also

enters

enters into a discussion of some historical facts not fully or accurately stated by Dr. Blackstone. This learned lawyer thought it necessary to make a reply, in which he declares that he had no view to Dr. Priestley in what he had said; that part of his Commentaries having been written fifteen years before; and that he was altogether unacquainted with his writings, his ingenious history of Electricity excepted. He openly disavowed the sentiment that "the spirit, the principles, and the practices of the sectaries are not calculated to make men good subjects;" and generously promised to cancel the offensive paragraphs in the future editions of his work. Dr. Priestley addressed a handsome and polite letter to Dr. Blackstone, in the St. James's Chronicle, which I remember to have read, either in that or some other newspaper at the time, and this brought the controversy, so far as Dr. Blackstone was concerned in it, to an amicable conclusion.

This controversy with Dr. Blackstone, led Dr. Priestley to write another pamphlet, entitled, *A View of the principles and conduct of the Protestant Dissenters, with respect to the civil and ecclesiastical constitution of England*, London, 1769. In this tract, after

after some general observations, he states particularly the religious principles of the Dissenters, and their objections against the constitution of the church of England; as claiming a power to decree rites and ceremonies; as establishing a *hierarchy*, consisting of orders of men, with titles and powers, absolutely unknown in the New Testament, &c.; on account of the practice of some useless and superstitious ceremonies; on account of the obstinate adherence to a form of prayer that contains many exceptionable passages. Lastly, the rational Dissenters have a class of objections peculiar to themselves, founded on the disbelief of the doctrine of the Trinity, and other points asserted in the liturgy or articles of the church of England. These heads are enlarged upon, and exemplified with great spirit and propriety. He next enters into a detail of the political principles of the Dissenters, and shews that there is nothing in them unfriendly to monarchy or the civil constitution of England, or to render them unworthy of the patronage or protection of government; to which, as settled at the Revolution, they and their ancestors have been the firmest friends. The whole concludes with a summary view of the history of the Puritans and some miscellaneous observations. Upon the whole, this is a very valuable performance, clearly and elegantly written, and highly worthy of the attention of that respectable body of men in whose favour it was composed.

The spirited tract above mentioned, was soon followed by another piece of a practical and sentimental nature, stiled, *A Free Address to Protestant Dissenters, as such.* By a Dissenter. The first edition of this piece was published at London, in 1769; the second, with enlargements, in 1771; and the third at Birmingham, in 1788. The two first were published without the author's name. In the preface, he assigns a very handsome reason for this concealment. "If it be asked, why the author chose to conceal his name, he frankly acknowledges, that it was not because he was afraid of making himself obnoxious to the members of the church of England. If they understand him right, they will perceive that his intentions towards them are far from being unfriendly; and if they understand him wrong, and put an unfair and uncandid construction upon what he has written, he trusts that, with a good meaning, and in a good cause, he will never be over-awed by the fear

of

of any thing that men may *think* of him, or *do* to him.

"Neither was it because he was apprehensive of giving offence, either to the *minister*, or to the *people*, among the dissenters, because he has spoken with equal freedom to both; but in reality, because he was unwilling to lessen the weight of his observations and advice, by any reflections that might be made on the persons from whom they come. An anonymous author is like the abstract idea of a man, which may be conceived to be as perfect as the imagination of the reader can make it.

"If, however, notwithstanding all the author's precautions, any of his readers should find him out, he hopes that, along with so much *sagacity*, they will at least have the goodness to forgive what was well intended, and excuse imperfections in one who is, at least, desirous to render others free from them."

After an animated exordium, the author treats in the first section *of the importance of the dissenting interest with respect to religion.* Under this head he shews, that it is only from dissenters that a reformation can be expected of those gross corruptions that have been introduced into religion; that princes and statesmen

statesmen only make use of it as an engine of state policy to promote their own secular ends; that all the service they can do to religion is not to intermeddle with it at all, so as to interrupt the reformations that might take place in it from natural and proper causes, &c.

"The kingdom of Christ (says Dr. Priestley) is not represented by any part of the metallic image of king Nebuchadnezzar, which denoted all the empires of this world; but is the *little stone cut out of the mountain without hands.* It is a thing quite *foreign* to the image, and will at last fall upon it and destroy all the remains of it. All that true christianity wishes, is to be unmolested by the kings and rulers of the earth, but it can never submit to their regulations.

"No christian prince before the reformation ever interfered in the business of religion, without establishing the abuses which had crept into it; and all that christian princes have done since the reformation, has tended to retard that great work; and to them and their interference it is manifestly owing, that it is no farther advanced at this day."

The

The reformation proposed by Wickliffe, so early as the year 1460, is shewn to have been more complete than any that has actually taken place in the church of England by the authority of the legislature. Errors and abuses have since been discovered, which Wickliffe did not suspect, but which affect the very vital parts of the christian system, and while adhered to, form an insuperable obstacle to its progress, such as the doctrine of the Trinity, &c. In order to remedy which abuses, the clergy must throw up their preferments, and the laity refuse to attend the established worship, in which case a reformation of the greatest abuses would immediately take place. Dissenters, in the mean time, ought to act the part that their situation enables them to do, by a rigid scrutiny into the foundation of their religious principles, rectifying what they find amiss, and using their endeavours to enlighten the minds of others. They ought conscientiously to forbear giving any countenance by a stated attendance on worship, that they believe to be unscriptural and idolatrous, which countenance on their part must have a natural tendency to perpetuate error and promote the cause of infidelity.

In section second, *the importance of the dissenting interest, with respect to the civil interests of the community*, is considered. The narrow views of the old Puritans with respect to civil and religious liberty, is candidly acknowledged and contrasted with the liberal ideas of their descendants, the present race of Dissenters. The just claims of this part of the community to a full participation of civil privileges are asserted, and at the same time they are consoled by truly christian motives and considerations, and exhorted to patience and acquiescence in the view of being deprived of them.

Section third, treats *of the manner in which Dissenters ought to speak or write concerning the church of England.* And here they are exhorted to integrity, and the most manly and open acknowledgment and profession of their sentiments respecting the divine unity and other important points. The lukewarmness and indifference, which the author saw with regret, growing up and spreading among the Dissenters of his time, founded either on false ideas of toleration and religious liberty, or arising from a sinful conformity to the fashionable world, are here severely

severely and deservedly censured, and a more strict and laudable conduct earnestly recommended.

The fourth section contains *observations on the expence attending the dissenting interest.* By the example of the primitive christians, and that of their ancestors the Puritans, the Dissenters are here exhorted to liberality in support of a good cause, which can never be maintained at too great an expence while it is considered as the cause of God and truth.

The fifth section gives excellent advice to ministers, with respect to their public and private conduct, manner of life, method of preaching, and discharge of their professional duty, highly deserving their most serious and attentive consideration.

Sections sixth and seventh, treats of the low and divided state of the dissenting interest, and the causes of it, which is shewn to be no just cause for abandoning it, but on the contrary to furnish motives for greater zeal and exertion.

P. 109. "The cause of *truth and liberty* can never cease to be respectable, whether its advocates be few or many. Rather, if the cause be just and honourable, the smaller is the party that support it, the fewer there are to share that honour with us. It

can

can never be matter of praise to any man to join a multitude, but to be singular in a good thing is the greatest praise. It shews a power of discernment, and fortitude of mind, not to be overborne by those unworthy motives, which are always on the side of the majority, whether their cause be good or bad."

P. 122. "Though it happens, that in the town in which you live, there be no society of Dissenters that you can entirely approve of, it can hardly happen but that there will be some, which, if you consider seriously, you may more conscientiously join with, than with the church of England. If we take in every thing relating to doctrine, discipline, and method of worship, I think there is no sect or denomination among us, that is not nearer to the standard of the gospel than the established church; so that, even in those circumstances, you will be a dissenter, if reason, and not passion or prejudice, be your guide.

"If when you reside for any time in the country, you chuse to go to church rather than to the dissenting meeting-house, because the dissenters happen to make no great figure in the place; if you feel any thing like *shame*, upon seeing the external meanness

meanness of the interest, and secretly wish to have your connections with it concealed; conclude, that the spirit of this world has got too much hold of you, and that religious motives have lost their influence.

"If this be your general practice (and I wish I could say it was not so, with many of the more opulent among us) you are but half a dissenter; and a few more worldly considerations would throw you entirely into the church of England, or into any other church upon earth. With this temper of mind you would, in primitive times, have been ashamed of *christianity* itself, and have joined the more fashionable and pompous heathen worship. But consider what our Lord says with a view to all such circumstances as these, *Whosoever shall be ashamed of me, and of my words, of him also shall the son of man be ashamed, when he cometh in the glory of his father, with his holy angels.*"

A postscript is added to this excellent address, in which dissenters are exhorted to a serious observation of the Lord's day, a regular attendance on public worship, and a proper concern to promote the cause of religion in the world.

The

The friendly care of our author to serve the cause of religion among the Dissenters, appeared soon after by the publication of another seasonable and valuable treatise, entitled, *A Free Address to Protestant Dissenters, on the subject of Church Discipline; with a Preliminary Discourse, concerning the Spirit of Christianity, and the corruption of it by false notions of religion*, London, 1770.

A sprightly animated vein of thought runs through this preliminary discourse; particularly that part where the love of Christ is considered as exciting a stronger sympathy, and having a more powerful practical effect, when he is contemplated as a man who lived and died to promote and secure the final happiness of mankind, apart from those false ideas annexed to his character in the Athanasian and Arian schemes. The true glory and dignity of Christ is also asserted in significant and affecting terms. The pernicious effects of false notions of religion on the mind, and the distress and uneasiness occasioned by them, are displayed and deplored.

The treatise itself contains eight sections. The first treats of the state and effect of church discipline in primitive times, and opens with the following just description

description of the end and design of christianity in general, and the institution of christian societies in particular.

"We are sufficiently authorised to say, that the great end which the Almighty had in view, in the dispensation of the gospel, was the reformation of a sinful world; and that whereas before the coming of Christ, the Jewish nation, alone, was honoured with the title of the *peculiar people of God*, the general promulgation of the gospel of Christ was intended to procure him, from *all* nations promiscuously, *a peculiar people zealous of good works.*

"Every christian society, therefore, having the same object in a particular place, that christianity in general has with respect to the world at large, should be considered as a voluntary association of persons who embrace christianity, and who are desirous of taking the most effectual methods to promote the real ends of it; or, in the language of scripture, *to build themselves up in the holy faith whereof they have made profession, to edify one another, and to provoke unto love and good works.*

"The members of christian societies are to exhibit to the world around them, an engaging pattern of of christian virtue, faith, hope, and joy; that *others, seeing their good works, may glorify their father who is in heaven.*

" In every christian church, therefore, there should be provision for admonishing all those who transgress their duty; for *reproving*, *rebuking*, or *exhorting*, for taking every favourable opportunity of suggesting useful hints, cautions, and encouragements; in order to make good impressions on the minds of all, at those seasons in which they are most likely to be deep and lasting, as in time of sickness, affliction, and distress. More especially, there should be proper provision that children and youth be particularly attended to, that they be carefully instructed in the fundamental and practical principles of christianity, in order that they may be well prepared for entering upon life with advantage, and be proof against the temptations and snares to which they will be exposed in it. Lastly, the weak and wavering should be strengthened, and have their difficulties removed. By this means, the great motives to a holy life being continually kept in remembrance, every member of the society may be *prepared for every good word and work*, be disposed to act with

with propriety and dignity, as becomes men and christians, upon every occasion in life, and to die with composure and good hope."

Our author observes "that the plan of a christian church was originally the same with that of a Jewish synagogue. Synagogues were places set apart for the reading and expounding of the law, and also for prayer. Here the people in the neighbourhood assembled for these purposes, every sabbath-day. A number of the more elderly persons, and those who had the most influence in the neighbourhood, had the title of elders, were appointed rulers of the synagogue, and had some kind of authority over those who belonged to the place; and some one of them was generally distinguished from the rest, but only by precedence, and having the direction of the service. The apostles and primitive christians, having been used to these regulations in places of public worship, adopted them in the constitution of christian churches.

"When, therefore, in consequence of preaching the gospel in any place, a number of persons were converted, the apostles immediately formed them into a *regular body*, and appointed proper *officers*.

 Those

Those who were distinguished for their age, gravity, good character, and knowledge, were made *presbyters* or *elders;* or, as they were sometimes called, *bishops*, though the last title was very soon appropriated to one of them; who was not, however, superior to the other elders in rank or authority, but only (to prevent confusion) presided in the assembly, and superintended the business of preaching, baptizing, and administering the Lord's supper. He also gave orders with respect to some other things, in which a number could not act to advantage.

"Besides elders and bishops, *deacons* also were appointed. They were persons whose business it was to assist the elders and the bishop, particularly in administering to the poor, and in other things that were of a civil, and not of a spiritual nature.

"But it was a fundamental principle in the constitution of the primitive churches, that no regulation, or resolution, respecting the state of the whole church, could be made but by the body of the people. They also chose the bishop and the elders, as well as the deacons.

"Epiphanius, who flourished A. D. 360, says, that nothing was necessary to the regular constitution

tion of a church, but elders and deacons; and that in churches where none of the elders were thought worthy of any distinguished rank, there was no bishop.

"It was the business of the elders, and by no means of the bishop only (who, in this respect, was only considered as one of them) to watch over the society, for the moral and religious purposes above mentioned. This is very evident from the book of acts, and the apostolical epistles."

The view our author has given of the constitution of the first christian churches, is justified by express quotations from scripture and early ecclesiastical writers, and the state of church discipline, and the impartiality with which it was administered, is described.

Section second, exhibits an account of the corruption and decay of the primitive church discipline, arising from the introduction of diocesan episcopacy; by church censures having been employed to animadvert upon particular *opinions* as well as practices; by the annexing of civil penalties to the sentence of excommunication; and lastly, by the injunction of *penances*, some of which were of a scandalous and ridiculous

diculous nature, and the commutation of these for sums of money, &c.

Section third, gives an account of the low and imperfect state of church discipline among rational Dissenters, in which, remarks are introduced concerning the state of things with respect to this article, in the church of England, and among the Presbyterians and Independents.

In section fourth, the circumstances are related that have brought about the change described in the preceding. Section fifth treats of the original state, progress, and present estimation of preaching; and in section sixth, a delineation of a method of church government, coming pretty near to the primitive plan, is given; in treating of which many pious and edifying observations are introduced. In section seventh, objections to this scheme of church discipline are considered, and some of its advantages more distinctly pointed out; and section eighth, suggests additional considerations as motives to the establishment of it.

Upon the whole, this treatise on church discipline is one of the most valuable of Dr. Priestley's practical pieces, abounding in fine, moral, and instructive sentiments,

sentiments, highly worthy the attention of christian ministers and people, and calculated to have excellent effects upon the minds of all who retain a proper attachment to the purity of christian morals. To use the words of the author, p. 115, "Should any society of rational christians, despising the insignificant censures of the world, form themselves upon this model, having no other object than the genuine simplicity of christian doctrine, and the native purity of christian manners, they would do themselves immortal honour; and, should their example be generally followed, they might be said, in a manner, to re-christianize the world."

The various pieces that Dr. Priestley had published relating to the Dissenters, with his occasional attacks upon the church of England, brought upon him the censure of an anonymous writer, himself a Dissenter, to whom the Doctor replied in a Tract, with the following title, *Letter to the Author of Remarks on several late Publications relative to the Dissenters, in a Letter to Dr. Priestley*, London, 1770. In this Tract, consisting of twelve letters, a particular reply is given to the objections of this anonymous writer, the Doctor's former writings are vindicated,

particularly

particularly his *Free Address to Protestant Dissenters, as such.* The church of England is charged with idolatry, and the nature of Athanasian idolatry is considered, and other topics are treated of, which the anonymous censurer had led the Doctor to touch upon.

In the same year he published, *An Answer to a second Letter to Dr. Priestley,* dated Leeds, Sept. 6, 1770. In this short Tract, consisting of four pages, 8vo. close print, Dr. Priestley replies to several complaints and charges made against him by the author of the *Remarks,* &c. in answer to his former set of letters concerning the Dissenting Interest. He confines his former assertions with respect to the topics in discussion, censures the maxims of the writer as being of a lax and trimming cast, and insists that his charge of *idolatry* upon those who pay divine honours to Jesus Christ, is just and well founded.

About the year 1770, was first published, *An Appeal to the serious and candid Professors of Christianity, on the following subjects,* viz. 1. *The use of Reason in matters of Religion.* 2. *The Power of Man to do the will of God.* 3. *Original Sin.* 4. *Election and Reprobation.* 5. *The Divinity of Christ;*

Christ; 6. *Atonement for Sin, by the Death of Christ.* This little Tract has had a most extensive circulation in England, Scotland and Ireland, and is too well known to require any particular account to be given of it. It was written by the author with the humane and benevolent design of enlightening the minds of the common people. The fifth section, which treats of the unity of God in the person of the Father, and the true nature of Christ is particularly valuable. The scriptural quotations are well selected, and forcibly urged. The conclusion is pathetic, affecting, and edifying. The first editions were sold for *one penny* each copy. It was afterwards enlarged, with a concise history of the rise of the doctrines mentioned in it, and an account of the trial of Mr. Elwall, and sold for *threepence.* At the time, or soon after the Appeal was published, the trial of Mr. Elwall was re-printed separately, with some extracts from his other writings in the first edition, and afterwards with extracts from the Unitarian writings of William Penn, the celebrated founder of the state of Pennsylvania. The trial of Elwall was re-printed at Dundee, in Scotland, and sold for a *halfpenny.* A number of answers having appeared to the Appeal

soon

soon after its publication, Dr. Priestley published *A Familiar Illustration of certain passages of Scripture*, relating to the six points discussed in the Appeal; to which he added an excellent prayer respecting the present state of christianity. This piece was intended as a confirmation of the Appeal, and a reply to all who had animadverted upon it. In the conclusion, he expresses his views and expectations with respect to this, and the two other small pieces he had composed with a view to instruct the inferior ranks of mankind. A truly christian object, worthy of Dr. Priestley, but too often neglected by the Dissenters of this world, who write only for fame, emolument, or to maintain the spirit, power, or credit of a party.

About the same time, our Author published, *Considerations for the Use of Young Men, and the Parents of Young Men*. Price twopence.

In this piece, the evils attending the irregular indulgence of sensual appetites and desires, are laid before youth in a clear, convincing, and powerful manner, and the cultivation of the virtues of purity and chastity strongly enforced. Early marriage, even previous to the acquisition of a fortune, is recommended, as an incentive to industry, frugality, and other

other virtues. The whole is closed with pious reflections, in which some pertinent quotations from Scripture are introduced.

We now come to consider the largest and most important publication that came from Dr. Priestley's pen during his residence at Leeds, viz. *Institutes of Natural and Revealed Religion*, first printed in 3 vols. 12mo. coming forth soon after one another, and afterwards re-printed twice in 2 vols. 8vo. and 12mo. This work is dedicated to the younger part of the congregation of Protestant Dissenters, at Mill-hill, in Leeds. The dedication bears date Leeds, March, 1772; and was afterwards recommended by the author to the consideration of the younger part of his congregation at Birmingham, Jan. 1, 1782. It was drawn up at first when the Author attended the academy, but had no doubt received accessions and improvements during the space of time that intervened before its publication. This work is extremely well calculated for the perusal of young people, and was used by the author as a foundation for lectures for that purpose. It is none of the least of its recommendations, that abstruse and sublime subjects are treated in it with such a degree of perspicuity as to render them intelligible to persons of ordinary apprehension. To the work is prefixed an instructive *Essay, on the best method of communicating Christian Knowledge to the Members of Christian Societies.*

The first part is divided into three chapters, relating to the being and attributes of God, and the duty and future expectations of mankind; and these three chapters are judiciously subdivided into several sections, in which, whatever can be inferred con-concerning the divine character, the passions and affections of men, their relations to one another and future prospects, from nature's unassisted light, or the proper exercise of our rational faculties, is distinctly stated and defined. Though the author has thought proper to guard what he has advanced on this part of his subject with the following necessary caveat: "Such are the conclusions which nature teaches, or rather which she *assents to*, concerning the nature and perfections of God, the rule of human duty, and the future expectations of mankind. I say *assents to*, because, if we examine the actual state of this kind of knowledge, in any part of the world, not enlightened by revelation, we shall find their ideas of God, of virtue, and of a future state, to have been

very

very lame and imperfect, as will be shewn more particularly when we consider, in the next part of this course, the *want* and the *evidence* of DIVINE REVELATION."

In the second part, after a sensible and spirited introduction, our author shews the origin and corruption of natural religion in general, and of *theology* in particular, the progress of idolatry, and the shocking superstitions that abounded in the heathen world, the imperfect conceptions that the philosophers entertained concerning God, the moral sentiments of the heathens, and their notions of a future life.

From the wretched state of morals and theology in the heathen world, and the deplorable circumstances in which mankind were placed, the probability of a divine interference is very justly inferred. Our Author, therefore, proceeds to state the positive evidences of revelation, and, previous thereto, he considers the nature, use, and credibility of miracles, the importance of testimony, with rules for estimating its value; he then opens the *antecedent credibility* of the Jewish and Christian revelations, the authenticity of the books of scripture, the evidence from testimony in favour of the christian revelation,

the

the evidence from the *resurrection* of Christ, and other facts of a similar nature, and the credibility of the Old Testament history.

He displays the evidence of the Jewish and Christian revelations from present appearances, viz. from their existence, propagation, and good effects, from standing customs, and internal marks of truth.

Lastly, he states the evidence of revealed religion from prophecies relating to *various nations* which had connections with the Jews, viz. *Ishmael* and his posterity, the *Arabs*, *Nineveh*, *Babylon*, *Tyre*, and *Egypt:* from prophecies relating to the Messiah, viz. Gen. xlix. 10. Is. xi. 1. Ps. ii. 7, 8. Is. xlix. 6. Jer. xxiii. 5. xxiii. 14. Micah v. 2. Is. ix. 1. lii. 13. liii. 1, &c. Zech. ix. 9. Hag. ii. 4. Dan. ix. 24. Is. xi. 1, 2, 3. Mat. iii. 1, &c. iv. 2, 5, 6; from the prophecies in the New Testament relative to the destruction of Jerusalem, the corruption of christianity, and the rise of the antichristian power mentioned by Paul, and John in the Revelation.

He next examines the pretended miracles of Apollonius Tyanæus, the magical rites of the heathens, the Popish miracles ascribed to the Abbe Paris, and one that was said to have been wrought

among

among the Camisards, and shews from the circumstances attending them, or the account given of them, that they are destitute of credibility, and cannot be fairly urged to invalidate the truth and evidence of the miracles which have been wrought for the confirmation of the Jewish and Christian revelations. Lastly, he replies to various objections against the Old and New Testament, and the facts and doctrines contained in them.

To this part of the work there is an Appendix, containing an ingenious *Essay on the analogy between the methods by which the perfection and happiness of men are promoted, according to the dispensations of natural and revealed Religion*, re-printed from the Theological Repository.

The third part, treats of the Doctrines of Revelation. In the introduction, the Author makes some remarks on the writings of Doctors Reid, Oswald, and Beattie: and in chapter first, after some observations on the unity of God, and the unhappy departure of the generality of mankind from this important tenet, the basis of the Jewish and Christian revelations, he recites a great many passages from Scriptures, well selected, relating to the unity, natural

tural perfections, and providence of God, and also his moral perfections. It is impossible to read this detail of quotations, without being struck with the just and sublime sentiments of the sacred writers, and acknowledging their vast superiority to the most admired writings of Pagan antiquity. The various branches of piety and moral duty towards the Creator, his creatures, and ourselves, are stated in the words of scripture, with occasional remarks. The positive institutions of revelation follow next in order, viz. the Observance of the Sabbath, Sacrifices, the Jewish Ritual, Baptism, the Lord's Supper, and the Government of Christian Churches. Lastly, a future state in general, the nature of future rewards and punishments, the duration of the latter, with the future condition of the world, are treated of. The work closes with an Appendix, in two sections, concerning other intelligent beings besides man, and abstinence from blood. This Treatise has been very properly adopted by the *Unitarian Society of Great Britain for promoting Christian Knowledge and the Practice of Virtue*, into the number of their books, and is sold at the very moderate price of 3*s*. 6*d*. in boards.

boards. By this means, its general circulation will be secured.

Before we take our final farewell of Dr. Priestley at Leeds, we must go back a little, and offer some observations on a celebrated and valuable work projected by him and carried into execution, with the assistance of others, during his residence at that place, entitled, *The Theological Repository*. The proposals for this work bear date *Leeds*, Nov. 1768. The persons who concurred with Dr. Priestley in this undertaking, were Mr. Cappe of York, Mr. Clarke of Birmingham, Dr. Kippis, Mr. Merivale of Exeter, and Mr. Turner of Wakefield. The plan was approved of by several other persons, and particularly by Mr. Aikin of Warrington, and Dr. Price. The chief burden, however, of conducting and arranging the whole, lay upon Dr. Priestley, and he received no assistance from any of the names before mentioned, except his near neighbour, Mr. Turner, of Wakefield. (See advertisement at the end of vol. 1.)

Many valuable original essays on various theological topics and critical disquisitions, made their appearance in this Repository; the three first volumes

were

were finished in 1771, and then the work was discontinued for several years. Mr. Turner of Wakefield, Mr. Crodale, Mr. Merivale of Exeter, Mr. Lindsey, and others, contributed to enrich the work with their ingenious productions. But none made a more capital figure than Dr. Priestley himself. The signatures he assumed, were *Clemens*, *Liberius*, and *Paulinus*; and the pieces that were written under these signatures, were the following, viz.

1. The one great end of the life and death of Christ; or the death of Christ no sacrifice or satisfaction for sin. Vol. 1, p. 17, 121, 195, 247, 327, 400.

2. Observations and queries concerning Judas Iscariot's being present, or not present, at the institution of the Lord's supper. Vol. 1, p. 141.

3. Observations on Christ's proof of a resurrection, from the book of Moses. Vol. 1, p. 300.

4. Observations on the apostleship of Mathias. Vol. 1, p. 376.

5. Essays on the Harmony of the Evangelists. Vol. 2, p. 38, 98, 230, 313.

6. Remarks on Rom. v. 12-14. Vol. 2, p. 154.

7. Observations

7. Observations on St. Paul's reasoning concerning Melchizedec. Vol. 2, p. 283.

8. Observations on the Abrahamic covenant. Vol. 2, p. 396.

9. An Essay on the Analogy there is between the methods by which the perfection and happiness of men are promoted, according to the dispensations of natural and revealed religion. Vol. 3, p. 4.

10. Observations on the reasoning of St. Paul. Vol. 3, p. 87, 188.

11. Observations on Infant Baptism. Vol. 3, p. 231.

12. An Essay on the Importance of Faith in Christ. Vol. 3, p. 239.

13. A Criticism on 1 Cor. xv. 27. Vol. 3, p. 255.

14. General arguments in favour of the Socinian Hypothesis, and an explanation of some texts which seem to be unfavourable to it. Vol. 3, p. 345, 357.

15. Observations on Christ's Agony in the Garden.

16. Observations on the Harmony of the Evangelists. Vol. 3, p. 462.

Leaving the curious objects discussed in these pieces to the consideration of the inquisitive reader, I shall only notice two of them, which I apprehend to be of more importance than the rest, viz. the first and fourteenth. This last mentioned, containing *General Arguments in favour of the Socinian Hypothesis*, &c. relates to a subject at that time much debated, and which underwent a particular discussion from different writers in the Repository itself. The general arguments here stated carry a great degree of weight in them in favour of the humanity of Christ, and against the Arian notion of his pre-existence, and several passages of scripture which seem at first sight to favour Arianism, and have often been urged for that purpose, are very ingeniously solved and explained, particularly John xvi. 28. and v. 13. Here I shall turn to the first Essay mentioned, which I consider as one of the most finished and elaborate that ever proceeded from the pen of Dr. Priestley, viz. *The one great end of the life and death of Christ*. In this Essay, every thing materlal in his *Scripture Doctrine of Remission*, published in 1761, is inserted, and the whole is much enlarged and improved.

He

He first endeavours, by a pretty long process of ingenious reasoning, to prove, that the principal and distinguished object of christianity, was to *ascertain* and *exemplify the important doctrine of a future state*, and that the other real objects and ends of the life and death of Christ, do all either flow from, or are perfectly consistent with this principal end.

Besides this primary end or object, he enumerates nine other dependent or subordinate ends.

2. If Christ lived and died to ascertain and exemplify the doctrine of a future state, and if, as has been represented, it was impossible that this should have been done without his actual death and resurrection, *he* certainly *died for us*, or on our account; and without his death, the great ends of his mission, our salvation from sin, could not not have been gained, which gives the greatest propriety to all such texts as the following: John x. 11-18 xv. 13. Rom. iv. 25. 1 Peter iii. 18. 3. Christ came to do the will of God. 4. To afford an example of voluntary obedience and suffering virtue. 5. He suffered and died to perfect his character. 6. To qualify him for obtaining a glorious reward, which might afford a strong motive of obedience to all his followers.

The

The rest are of such a nature that they cannot be conveniently abridged, and all are illustrated by texts of scripture quoted at full length.

Having thus stated his own ideas with respect to the life and death of Christ, he proceeds to oppose those notions that he considers as erroneous and ill-founded. After some remarks on the figurative and metaphorical style of oriental nations, and particularly of the sacred writers, and the mistakes that have arisen from too literal an interpretation; he quotes several passages at full length, in which Christ is represented as a sacrifice, either expressly, or by plain reference, viz. John i. 29. Eph. v. 2. Heb. 7. 27, and refers to various other passages in the same epistle, 1 Pet. i. 2-18. 1 John ii. 2. iv. 10. Rev. v. 6-9. Is. liii. 10. 2 Cor. 5. 21. He refers to Heb. i. 3. vii. 25. ix. 11. xvi. 12, 14, 21, where he is called *a priest*, and *a high priest*. He considers the language of these passages as figurative, so far as respects the terms sacrifice and priest, and assigns six distinct arguments to justify this method of interpretation. The second of these is, " that the Jewish sacrifices are no where said, in the Old Testament, to have any reference to another more perfect sacrifice, as might have

have been expected, if they really had referred to any such more perfect sacrifice, and such an one had been necesssry. On the contrary, whenever the legal sacrifices are declared, by the prophets, to be insufficient to procure the favour of God, the only thing that is opposed to them as of more value in the sight of God, is personal holiness, good works, or moral virtue." Under this, he quotes the following texts at length, Psal. li. 16, 17. Is. i. 11-20. Hos. vi. 6. Amos v. 22-27. Micah vi. 6, 7, 8. Mark xii. 32, &c.

He inclines to think that sacrifices were not, in their origin, of divine institution, but took their rise from the corporeal ideas men had of God in early ages in the infancy of the world, and were offered as *gifts*, *presents*, *entertainments*, or circumstances accompanying an address to the Deity, Psal. i. 8, &c. "It was not the *sacrifice*, but the priest that was said, in the Old Testament, to make *atonement*. Nor was a sacrifice universally necessary for that purpose; for, upon several occasions, we read of atonement being made, when there was no sacrifice. Thus Phinehas is said to have made atonement for the children of Israel, by slaying the transgressors, Num.

Num. xxv. 13. Moses made atonement by prayer only, Exod. xxxii. 30. and Aaron made atonement with incense, Num. xvi. 46, 47.

He differs from the author of *Jesus Christ the Mediator*, who says, "that in the very notion of sacrifice there was respect to sin;" and opposes the sentiments of Dr. Taylor, that sacrifices were *a symbolical address to God*, &c. He quotes the texts in which the term λυτρον, *ransom*, is used in the New Testament, or which convey a similar meaning, Math. xx. 28. Mark x. 45. John xi. 50. 1 Tim. ii. 8; and introduces the remark of Vigilius (Mr. Turner of Wakefield) upon some of these texts, who summing up what he has advanced upon them, observes, "Upon the whole, it appears, that where-ever any of the derivations from λυτρον, are applied to Christ, and especially to his death, they convey no idea of a price paid to ransom men from the penalties of the broken law, but of a moral expedient to deliver men from subjection to, and the practice of sin; and I think this is the precise meaning of λυτρον in the texts of Mathew and Mark." The same learned writer shews, that though the preposition αντι, sometimes signifies *instead of*, yet in various places it signifies,

nifies, *because of*, or *therefore*, as Luke i. 20. ii. 3. xix. 44. Acts xii. 23. Eph. 5. 31. Heb. xii. 2, &c. He treats of the texts which represent Christ as *bearing the sins of mankind*, Isa. liii. 11, 12. 1 Pet. ii. 34. Heb. ix. 28; and explains the term *bear* in the sense of *bearing away* or removing, and thinks his understanding them in this sense justified by John i. 29. 1 John iii. 5-6. Math. viii. 17, and observes, that "the phrase *bearing sin*, is never applied under the law, but to the *scape-goat* on the day of expiation, which was not sacrificed, but, as the name expresses, was turned out into the wilderness, a place not inhabited."

He considers various representations of the death of Christ in the New Testament, viz. as a *curse*, Gal. iii. 10; as a *passover*, 1 Cor. 5, 7; as a *testament*, enforced by the death of the testator, as having a resemblance to *the serpent which was exposed by Moses in the wilderness*; and concludes from the various and opposite nature of the representations, that they are probably intended as figurative allusions and comparisons, and ought not to be taken in a gross and literal sense. He concludes this part of the subject by quoting Rom. vi. 3-6-8. Gal. ii. 20-24.

vi.

vi. 14. Eph. ii. 5, 6; in which the strongest figures derived from the sufferings, death, and resurrection of Christ, are applied by the apostle Paul to the state and condition of christians in general, and his own in particular, the greater part of which metaphors are bolder, and more far-fetched, than the comparison of the death of Christ to a sacrifice, and shew that the apostles were fond of considering it in a moral view, as affording the strongest and noblest motives to a holy life.

He produces several other passages from the Gospels and Epistles, as John xx. 31. Acts x. 43. xiii. 39. Rom. 3. 24. 1 Cor. vi. 11. Gal. ii. 16. Eph. iv. 32. 1 Thess. i. 10. Heb. vii. 25. 1 John ii. 1, &c.

After remarks on these passages, and others connected with them, he has the following observation: "Upon a careful examination of these and other texts, produced for the *commonly received* doctrine of atonement, it must be granted, that some do seem to represent the pardon of sin as dispensed in consideration of something else than our repentance or personal virtue; and according to their literal sense, the pardon of sin is, in some way or other, procured by Christ. And had the literal representation been

all

all of a piece; had the sacred writers uniformly represented God the Father as dispensing the pardon of sin to penitent offenders, in consideration either of the sufferings, or of the merit of Christ, our only intercessor, the account would have had more of the air and consistency, at least, of truth: But when the pardon of sin is sometimes represented as dispensed in consideration of the *sufferings*, sometimes of the *merit*, sometimes of the *resurrection*, and even of the *life* and *obedience* of Christ; when it is sometimes Christ, and sometimes the spirit that intercedes for us; when the dispensing of pardon is sometimes said to be the proper act of God the Father; and again, when it is Christ who forgives us, we can hardly hesitate in concluding, that these must be, severally, *partial representations*, in the nature of figures and allusions, which, at proper distances, are allowed to be inconsistent, without any charge of impropriety in the stile of the composition."

From these texts, the Author appeals to the general sense of scripture, in which the pardon of sin is represented as dispensed solely on account of men's personal virtue, a penitent upright heart, and a reformed exemplary life, without the least regard to

 the

the sufferings or merit of any being whatever. In proofs of which, the following passages are alleged: Gen. iv. 7. Psal. xv. throughout, li. 17. Is. iii. 10. lv. 7. Math. v. 1-12. Our Saviour's beautitudes, vi. 14. John xii. 16. Acts x. 34. ii. 37. iii. 19. xvii. 31. Rom. xiv. 17, 18. 1 Cor. xv. 58. 2 Cor. i. 12. 1 Tim. iv. 8. Jam. ii. 24. Rev. ii. 10. xiv. 13. xxii. 14.

The absolute declarations of mercy and favour to the penitent and virtuous, are quoted in the following places: Exod. xxxiv. 6-7. 2 Chron. xxx. 9. Psal. xxv. 8. lxxxvi. 5. ciii. 8. Is. xxx. 18. Ezek. xxxiii. 11-14, &c. Dan. ix. 3. Michah vii. 18. Jonah iv. 2. John xvi. 26. 2 Pet. iii. 9. 1 John i. 9. Rom. iii. 24. Titus iii. 7.

The penitential addresses of David, in which he pleads the free mercy of the Divine Being, and sometimes his own integrity, and the stress Hezekiah and the worthy Nehemiah laid upon good works, are exhibited in these places, Psal. vi. 4. xxv. 6. li. 1. vii. 8. Isa. xxxviii. 3. Nehem. v. 19. xiii. 14-22. Our author reasons strongly from these, and the foregoing passages, as utterly inconsistent with the popular doctrines of atonement and the inefficacy of good works,

works, and expresses his surprise "that, in all the books of scripture, neither in the Old or New Testament, neither the Divine Being himself, to the patriarchs; neither Moses, nor the prophets, by his direction, to the Jews; nor Christ, or his apostles, to the christians, ever assert, or explain, *the principle on which the doctrine of atonement is founded*; for though they describe the heinous nature of sin, in the strongest colours, represent it as exceeding sinful, and the like, they never once go a single step further, and assert that *it is of so heinous a nature, that God*, the infinitely good and gracious, *cannot pardon it, without an adequate satisfaction being made to his justice*, and the honour of his laws and government." The author of *Jesus Christ the Mediator*, having asserted, that the principles on which the doctrine of atonement is founded, are laid down by the apostle Paul in Rom. iii. 25-26, our author thinks that the passage, when rightly rendered, affords no foundation for such an assertion.

Our Author thinks, that if it had been the great end of Christ's coming into the world, to make satisfaction to the justice of God either for the sins of the whole world, or those of the elect only, we might expect

expect to find sufficient reference to it in the history and discourses of Christ, and also that the promised Messiah should have been *announced* before-hand by the Jewish prophets in this important light. But after a pretty exact scrutiny into the contents of the gospels, he does not find any such doctrine deliver-ed in them, either in the accounts we have of our Lord's birth, the declarations of John the Baptist concerning him, or his own discourses, Mark i. 14. These last are chiefly in a moral strain. "He in-veighs freely against all the prevailing vices and irre-gularities of his time, and mentions all the more ag-gravating circumstances of them; but he never hints at any satisfaction being made to the justice of God for them. He makes a fine encomium upon several moral virtues, and pronounces, absolutely, such and such characters to be fit for the kingdom of God, but never with any such cautions or restrictions as are generally given at this day, letting us understand, that these virtuous qualifications *alone* will not en-title a man to a place there."

"The sermon on the Mount contains excellent moral lessons, but nothing else. Without the least mention of any method of making the deity placable,

he

he pronounces, clearly and authoritatively, what characters were entitled to the blessings of the kingdom of heaven, and what were not entitled to them. He also says, Math. vii. 21. *Not every one that says unto me, Lord, Lord, shall enter into the kingdom of heaven, but he that doth the will of my father who is in heaven.* When a certain lawyer asked him, *What shall I do to inherit eternal life?* he said unto him, *What is written in the law, how readest thou?* And he answering said, *Thou shalt love the Lord thy God with all thy heart, and with all thy soul, and with all thy strength, and with all thy mind, and thy neighbour as thyself. And he said unto him, thou hast answered well; this do, and thou shalt live.* Those who maintain the commonly received doctrine of atonement, and the insignificance of every thing that fallen man can do, to obtain the favour of God, can make nothing of this passage, but by supposing that our Lord spoke here ironically, a supposition which my reader, I dare-say, will not ask me to refute."

"When a certain ruler asked him, saying, Math. xix. 16. Mark x. 17. Luke xviii. 18. Good master, what shall I do to inherit eternal life, he still refers them

them to the commandments, and bids him also *sell all that he had and follow him, assuring him that then he would have treasure in heaven.*

"When the woman was caught in adultery, he says to her, John viii. 11. *Neither do I condemn thee; go, sin no more.* And when Zaccheus made profession of his repentance, Luke xix. 9. *Jesus said unto him, This day is salvation come to this house; for the son of man is come to seek and to save that which was lost.*"

"In the representation that our Lord makes of the transactions of the day of judgment, in the 25th chapter of Matthew, there is no mention of any thing but of good or bad works. The righteous, agreeably to their character, think humbly of themselves, and will hardly believe that they have done any thing very meritorious: they are surprised and overcome with joy at the approbation of their virtue and merit, but never refer themselves to the sufferings, or to the merit of their advocate and judge, for the ground of their hopes; though nothing in the world can be conceived to have been more natural and pertinent upon the occasion."

"When

"When our Lord directs his disciples to pray for the forgiveness of sins, in that excellent form which is commonly called the Lord's Prayer, Matt. vi. 12. doth he give the most distant hint of the pardon of sin being dispensed in consideration of what *he* should do or suffer for them? On this occasion, he surely could not have omitted representing himself in this light, if the sins of men had really been forgiven on his account; and especially, if a regard to his death or merit, had been necessary to the obtaining the remission of sins. The form is nothing more than this: *Forgive us our trespasses, as we forgive them who trespass against us.* And so far is he from giving a hint of any deficiency in this form, that what he subjoins, by way of explication, with respect to this most important petition, is as clear a confutation of the doctrine of atonement, as could be given by a person who had never heard of it, and could not suspect it. For he says, v. 14. *If ye forgive men their trespasses, your heavenly father will also forgive you. But if ye forgive not men their trespasses, neither will your father forgive your trespasses.* The same sentiment occurs, Math. xviii. 35. Mark xi. 25, &c."

Dr.

Dr. Priestley refers to our Lord's parable of the person who owed ten thousand talents, Math. xviii. 23. to the creditor who had two debtors, Luke xv. 18. &c. as exhibiting the sentiments and conduct of his heavenly Father with respect to the forgiveness of sin; and treats of various other passages in the gospels. He considers the omission of the commonly received doctrine of atonement in our Lord's solemn prayer, John xvii. in the history of his agony in the garden; in his trial before the Jewish Sanhedrim, Pilate and Herod, at his crucifixion, and after his resurrection, in the charge he gave to his disciples, to preach the gospel to all nations, as affording strong suspicions that this tenet is not an original doctrine of christianity, but has come into it in some such manner as other corruptions have been introduced. He appeals to Peter's discourse to the Jews, Acts ii. 33. iii. 17, 18. Stephen's apology, Acts vii. the conversation that took place between Philip and the Eunuch respecting Isa. liii. Acts viii. Peter's discourse to Cornelius, Acts x. 34, &c. Paul's discourses before the Jews at Antioch, Acts xiii. 28. at Thessalonica, chap. xvii. before Agrippa, chap. xxvi. and at Rome, chap. xxviii. to the heathens at Lystra,

Lystra, Acts xiv. and at Athens, chap. xvii. and sums up his observations upon these passages with the following remark; "When we find the apostles so absolutely silent, where, we cannot but think, there was the greatest occasion to open themselves freely concerning the doctrine of atonement; when, in their most serious discourses, they express themselves in language that really sets it aside; when they never once directly assert the necessity of any satisfaction for sin, or the insufficiency of our good works alone to entitle us to the favour of God and future happiness, must we build so important an article of our faith upon mere hints and inferences from their writings? The doctrine is of too much importance to be built on such a foundation."

Doctor Priestley proceeds to combat an objection that has been started against his scheme of divinity, viz. "that the apprehension of some farther satisfaction being made to divine justice than repentance and reformation, is necessary to allay the fear of sincere penitents." He asserts, "that it appears from the history of the opinions of mankind, that all men naturally apprehend the Deity to be propitious." In proof of this assertion, he considers the state of the patriarchal religion in the time of Job, and quotes the following passages at length: Job xi. 14, &c. xxii. 21, &c. xxxiv. 31. xxxv. 8, &c. v. 16. xlii. 5. He appeals to the history of the repentance of the Ninevites, Jonah iii. 4, &c. iv. 2. He recites some passages from the books called *Apocryphal*, *Wisdom* ix. 23. *Ecclesiasticus* xxxv. i. *Song of the Three Children* v. 12-16. He refers to the prayer of *Manasses*, and the history of the *Mother and her Sons*, mentioned in the book of Maccabees; and quotes Philo, Josephus, Maimonides, and other Jewish writers. He quotes Dr. Hyde and Grosse's voyage, for an account of the notions of the ancient *Persians* and modern *Persees;* and Mr. Dow's history of Indostan, for the ideas of the Indian Brachmans; and Le Page du Pratz, for those of the tribes of America."

Doctor Priestley denies that any satisfaction is necessary to the justice of God for sin: "For divine justice is not that blind principle, which, upon any provocation, craves satisfaction indiscriminately of all that comes within its reach, or that throw themselves in its way; but justice in the Deity can be no more than a modification of that *goodness*, or benevolence, which is his sole governing principle; the object

ject and end of which, is the supreme happiness of his creatures and subjects. His happiness being of a moral nature, must be chiefly promoted by such a constitution of the moral government we are under, as shall afford the most effectual motives to induce men to regulate their lives well. Every degree of severity, therefore, that is so circumstanced as not to have this tendency, *viz.* to promote repentance, and the practice of virtue, must be rejected by the benevolent principle of the moral government of God, as disagreeable even to divine justice, if it have the same end as the divine goodness, the happiness of God's creatures." He considers the doctrine of atonement in a practical view, and thinks the belief and influence of it unfavourable to virtue and morals.

Doctor Priestley concludes the whole of this ingenious Treatise, with an account of the scheme of salvation by Jesus Christ, according to his own conceptions of it, from which I shall give the following extract.

" I am very sensible that, after an attempt to shake the credit of a doctrine, which many persons look upon to be the most essential to christianity, as the most

most fundamental principle, the life and soul of the whole scheme; without which, all the rest is a mere dead, lifeless thing, destitute of spirit or meaning; the advocates for the doctrine of atonement will be ready to ask, what, if we give up this point, must be our notions of christianity? Wherein shall we differ from the *Deists?* Instead of making a direct reply to these queries, I shall subjoin, by way of conclusion, a concise view of the scheme of salvation by Jesus Christ, without the doctrine of atonement for sin.

"Let us, then, suppose the whole race of mankind to be in a state of apostacy from God, lost to all sense of religion and virtue, in the expressive language of scripture, *dead in trespasses and sins;* and that without a revelation and a saviour, they were in the most deplorably vicious and wretched circumstances, in a sure way to make themselves miserable both here and hereafter.

"In this state of things, God, the ever benevolent, who is *good to the unthankful and the unworthy*, wishes their happiness; but, rational and moral agents, as men are, cannot be made happy without being recovered to a sense and practice of their duty, which must, from its own nature, be a voluntary thing.

thing. What, therefore, can the ever blessed God, tender of our happiness, do for us? To force our compliance, would not answer the purpose. We must be *won upon*, be engaged by proper *motives* and considerations, to reform our hearts and our lives. Such measures must be taken with men as are suited to the nature of *reasonable beings*, and, at the same time, governed very much by views of interest, for such creatures are men.

"Now all that we conceive could possibly be done for us; all that even infinite wisdom, goodness, and power could contrive and execute, in order to our recovery, due regard being had to our nature, may be reduced to these following particulars. *First*, to instruct us in the whole of our duty. *Secondly*, to engage us to the performance of it, by the promise of suitable and sufficient rewards, and to deter us from disobedience by the fear of punishment. *Thirdly*, to draw us by a proper set of examples of virtue; and *lastly*, to give us the most satisfactory assurance of the pardon of our past sins upon our repentance and reformation, of the certain acceptance of our sincere, though imperfect endeavours to

do

do our duty, and of all necessary assistance in the practice of it.

"And hath not all this been done for us in the most ample manner, in a course of moral dispensations, commencing in the days of our first progenitors, and carried on with the utmost regularity, through the hands of the *patriarchs*, *Moses*, and the *prophets*, till it received its highest perfection by the hands of the blessed *son of God* himself? who, on this truly great and generous errand, submitted to all the infirmities, indignation, and inconveniencies of human life; and, to close the whole in the most advantageous manner, died a most painful and ignominious death."

"Is not here a scheme of salvation and redemption, in every part complete, without any atonement? Simple as it is, do not the few parts of which it consists, contain every thing that could be applied, with effect, for our benefit? Would not then any addition to it greatly clog and embarrass the scheme, and spoil its effect? You say, this leaves us at a loss to know what provision is made for the pardon of our sins; but what doth that concern us? If we have the fullest assurance, from the mouth of God

himself,

himself, that our sins are *actually* forgiven, upon our repentance, (though we did not know for what reason, or whether any other reason than repentance were necessary) should not that satisfy us? Is not this assurance, all that can possibly be any inducement to us to forsake our evil ways, and return to God and our duty?"

"You still ask, what necessity for the death of Christ upon this scheme? If he did not die to make satisfaction for our sins, must he not have died for some end that was very low, and unworthy of him? I reply, (by recapitulating what has been advanced before) is to die a martyr to the truth, to prove his divine mission in the most illustrious manner that we can conceive; to ascertain the reality of a resurrection, and a future life, to such beings as we ourselves are; to evidence the benevolence of his heart, and the greatness of his soul, the vast importance of the work he undertook, and how much his heart was in it; to encourage all who should hereafter embrace his religion, to lay down their lives with courage and cheerfulness, in the cause of truth and integrity, by giving an example of suffering virtue in his own agony and death; and that God, by his exaltation to glory,

glory, in reward of his obedience unto death, might thereby exhibit, before all who believe in him, an example of the certainty and greatness of the rewards, which he will confer on all who shall, in like manner, obey him ; were these and the other subordinate ends mentioned in the former part of this article, I say, low and unworthy of Christ ? And when we say that he died for these purposes, though we add no other, do we say that he died in vain ? When his death so circumstanced, looked with so friendly an aspect upon human virtue and happiness ; and when by this means, our Lord put the finishing hand to so extensive a scheme, in which was done whatever was practicable, to recover fallen man to immortal virtue and happiness, is he not with great propriety stiled, our *redeemer*, *saviour*, and *mediator ?*"

" And when in the word of God, we are taught to consider all the evils that infest this present world ; the laborious cultivation of the earth ; the shortness and infirmities of human life ; with death, and all the evils we can name, as the consequence of the introduction of sin into the world ; when the Almighty threatens impenitent sinners with unspeakable torments in the world to come ; when he hath put in execution

execution a scheme so astonishingly glorious and expensive, to redeem us from all iniquity; having given up his only son to die, in order to effect it; can we have any pretence for saying, that God hath not sufficiently testified his abhorrence of sin? What could he have done more, consistent with his perfections, and with the natures he had given us to testify that abhorrence.

"With this great, but simple scheme of religion, the apostles were entrusted, that they might publish it for the benefit of the world. And when we consider what vessels they were that it was deposited in, and to what immediate use it was to be applied, we shall not wonder at the tincture it received from them. The apostles were *Jews*, and they had to do with Jews. The phrases belonging to the Jewish religion were the most familiar to them, and the fittest, in the world, to make the simple doctrines of christianity take with their countrymen. At a profuse expence, therefore, of figures and allusions fetched from the Jewish ritual, to make the new religion the better to tally with the old; liberties too great for our *European* manners, but not greater than the Jewish nation had been accustomed to; at the experee, therefore, of no sincerity, or integrity, they suit their entertainment to the taste of those who were first to be invited to partake of it."

Such are the sentiments of Dr. Priestley with respect to the *one great end of the life and death of Christ*, the subordinate and secondary ends, and the doctrine of atonement, which I have endeavoured to exhibit with precision and candour. He has said enough, I apprehend, to establish the important and comfortable tenet of the placability of the divine nature, and to confute the Calvinistic doctrine of the necessity of a plenary satisfaction to the justice of God for sin: but whether his arguments be sufficient to overturn all the more moderate schemes that have been proposed with respect to this much litigated subject, I shall not pretend to affirm. The subject is viewed in a different light by some writers in the Theological Repository itself. See an *Essay* on the Sacrifice of Christ, Vol. 1, p. 173 to 183, and p. 225 to 236, by Theophilus. An Essay on praying in the name of Christ, Vol. 1, p. 363 to 376; and Observations on the Sacrifice of Christ, Vol. 2, p. 3 to 22, by Verus. Eusebius (Mr. Turner of Wakefield) published a Dissertation on the

meaning

meaning of atonement in the Old and New Testament, Vol. 3, p. 385 to 433. His method of treating the subject resembles that of Dr. Priestley more than that of any of the beforementioned writers; but his notions in all respects do not appear to me to be precisely the same. He observes, p. 431, "Thus I have taken notice of, and endeavoured to explain all the passages of the New Testament where I find the words *ιλασκομαι ιλασμος*, and *ιλαστηριον*, made use of and applied to Christ, and I apprehend that they all relate only to the establishment and confirmation of those advantages we at present enjoy by the gospel, and particularly of a free and uninterrupted liberty of worshipping God according to the institutions of Christ, granted unto us in consequence of his death; just as the legal atonements served (though far more imperfectly) similar purposes under that dispensation." And p. 433, after having quoted several passages in the New Testament, he adds, "In all these passages the death of Christ is represented as the *means* by which the reconciliation, redemption, or deliverance of mankind is effected, and the forgiveness of sins already conferred on believers, through the free grace of God, in order to their

their improvement in holiness, by the influences of the example, doctrine, and institutions of Christ, which are provided for the benefit of the whole community of his church. But I do not recollect any text, where the death of Christ is respresented, as the *cause*, *reason*, or *motive*, why God has conferred these blessings on men." See also, p. 425 to 429, too long to be here quoted.

In May 1773, Dr. Priestley took his farewell of his congregation at Leeds, in a pious and energetic discourse, from 1 Pet. i. 13, which was afterwards published. Some months before his resignation of the pastoral office, he had declared his intention to the congregation by letter, and received from them a reply expressive of their affectionate esteem for his person, and their grateful sense of his sincere and faithful services as a minister, and earnest endeavours to instil knowledge and inculcate good principles on the minds of their children, and bearing testimony to the harmony, peace and friendship that had mutually prevailed during their connection. From Leeds, he went to reside with Lord Shelburne, in the capacity of a literary companion, and did not fail to

employ

employ the leisure he enjoyed in the composition of various literary works.

In the year 1772, a considerable number of respectable clergymen of the church of England, joined with some professors of medicine and law, united in an application to Parliament for relief from the obligation of subscription to the thirty-nine articles of the church of England. This petition, after having been debated in an animated manner in the House of Commons, was rejected by a majority of its members. The Dissenters also applied about the same time for a redress of their grievances, and to be relieved from an obligation to subscribe the doctrinal articles of the church of England, most unjustly imposed upon them by the legislature. But their application was unsuccessful at that time. Though Dr. Priestley took no active concern in these proceedings, for wise and good reasons, he could not remain an unconcerned spectator in a case that so nearly interested the Protestant Dissenters. He therefore published a pretty large pamphlet, with the following title, *A Letter of Advice to those Dissenters who conduct the Application to Parliament for Relief from certain Penal Laws, with various Observations*

servations relating to similar subjects, London, 1773. In this publication, he applauds the conduct and steadiness of those who had managed the application to Parliament. He recommends to them an enlargement of their views; that they should rise in their demands, and make each succeeding application an improvement upon the former; that not contented with a redress of those grievances that merely affected themselves, they should take the case of all their dissenting brethren under consideration, and request a general abolition of all penal laws without exception, and particularly of the act of William and Mary, which affects Unitarians of every description, that make an open profession of their opinions. With a high degree of generosity, he recommends to the Dissenters to petition for a bill by which unbelievers shall be as much at liberty to attack, as themselves to defend, either christianity in general, or their particular opinions concerning it.

He hints at some advantageous alterations that might be made in the ecclesiastical establishment itself; though he leaves what improvements are most necessary or expedient, to such persons as the Candid Disquisitors, the author of the Confessional, and the

the late petitioners among the clergy, together with Mr. Wollaston and his friends, to set forth. Such persons as these (he observes) who themselves feel the grievances, are best able to explain and judge of them. Concerning other alterations, which are entirely of a civil nature, he considers political men as better judges than the clergy, such as the inequality of ecclesiastical benefices, the mode of provision by tythes, and the temporal power of the bishops, all of which he thinks might be rectified or changed for the better. Lastly, he thinks it a proper subject of inquiry for the politician, whether, considering the many abuses to which ecclesiastical establishments are liable, there be really any utility in them at all, and whether the very great expence which always attends them, might not be applied to a better purpose. In the remaining part of this pamphlet, the Author treats of the offence he has given to those Dissenters who have conformed to the church of England, particularly Dr. Dawson; of the objection that has been made to the declaration of a belief in the scriptures proposed in the late bill, and of the opposition made to it by some of those who are called Orthodox Dissenters.

Before

Before I leave this pamphlet, I cannot help producing a pretty long, but highly interesting quotation from it. Though Dr. Priestley, in various places of his writings, has charged the established worship of the church of England with idolatry, both before and after the date of this publication, particularly in his *Letters concerning the Dissenting Interest*, p. 17, 21, and his *Familiar Letters to the Inhabitants of Birmingham*, Letter xvii. p. 126; yet in none of these has he stated the charge so minutely, or brought it so fully and completely home, as in the following passage. In justice to his memory, therefore, and to those upon whom the imputation falls, it ought to be exhibited and laid anew before the public; more especially as it is extracted from a part of his writings, at present perhaps little read or attended to.

P. 39. "No single thing in my writing seems to have given more offence to the clergy of the church of England, than my calling the established church an *idolatrous* one. But I cannot help thinking that the friends of that establishment would have done much better, if, instead of using those expressions of surprise and horror, which are still resounding from

from all quarters upon this occasion, they had waited in silence till the first consternation had abated, and they had been capable of considering the charge with composure and attention. On the contrary, I have not found that a serious refutation of the charge, which I have often avowed and repeated, and which I now again avow and repeat, has been so much as attempted; except, strange as it may seem, by a Dissenter, to whose apology for the church of England I have replied. Certainly, however, if the charge can be proved to be false, the odium which must otherwise fall upon the church of England, will recoil upon myself. But if it be true, the violent exclamations of the bishop of Llandaff, and his friends, will only tend to make the accusation more notorious, and more effectual.

"I cannot say that I am at all surprised at the manner in which this charge of idolatry has been received. It is, indeed, of no trivial, but of a most serious nature, importing that the religious establishment of this country comes under the description of the great *antichristian system*, which was to prevail in the latter times, as well as the church of Rome, the leading characters of which are a corruption of the christian principles, and a depravation of its spirit, usurpation of the rights of God and of Christ, idolatry, (and consequently blasphemy) and persecution by the civil power.

"Indeed, almost every sect of christians that has had power, has been chargeable with the same enormities, and *so far*, and *so long*, they also have borne the *mark of the beast*; but, next to the church of Rome, no christian church has ever borne those marks so evidently, and so long. It is, therefore, my serious opinion, that in that utter destruction of all antichristian corruptions and usurpations in christianity, which is clearly predicted by the prophets, the church of England will not escape; but that the impiety and idolatry of her tenets, her avowal of a claim to power which belongs to Christ only, viz. *authority in controversies of faith*, and the righteous blood which she has shed, together with every unjust restraint which she has laid upon men for conscience sake, will *come up in remembrance before God*, in those *days of vengeance*, the near approach of which, I own, I am looking for.

"But, heavy and serious as this charge of idolatry is, the principles on which it is grounded are sufficiently,

ficiently obvious and intelligible, so that, if it can be refuted, the refutation must be very easy; and consequently every thing, besides argument, must be superfluous, at least previous to argument.

"The question is simply this: Is there only one God? Or are there more Gods than one? Or, to avoid all ambiguity, are there more *intelligent agents* than one, who are uncreated, having an existence independent of all other beings, and to whom, as omnipresent and omnipotent, prayers may with propriety be addressed?

"The writers of the Old and New Testament, and that great Being by whom these writers were inspired, not only answer this important question in the negative, but every where lay the greatest stress upon that negative. The first of all the commandments is, *Thou shalt have no other Gods but me.* Moses, and the rest of the prophets, repeat this great doctrine so frequently, that the establishment of it cannot be denied to have been the greatest object of that dispensation of religion. Our Lord Jesus Christ, *the man approved of God, by miracles, and wonders, and signs, which God did by him,* who received all his doctrine from God, and whom God

raised

raised from the dead, calls this same God his own *God and Father*. He expressly addresses him under the character of the *only true God*, and asserts, that he only is to be worshipped.

"The apostles uniformly speak the same language, acknowledging only *one God, even the Father, and one mediator, the man Christ Jesus;* and to adopt the contrary sentiment, and to conduct divine worship agreeably to it, by multiplying objects of worship, has always been termed *idolatry;* and being in the highest degree derogatory from the most essential rights of that God, who has solemnly declared that he will not give his glory to another, cannot be termed less than impious and blasphemous. It must be as much so as any opinion, and practice grounded upon it, can be. It is even impossible to suppose a case in which those terms can be applied with more propriety.

"The worship of different created beings makes no sensible difference in this respect, since an *archangel* and a *stone* are equally what God makes them to be. Their peculiar powers and properties are liable to be destroyed and changed at his pleasure; and with the same ease he can even annihilate them

both;

both; so that the *worship* of either of them, *as God*, is equally absurd and impious.

"Now, do the articles, and public offices of the church of England, uniformly speak the language of the scriptures concerning the proper unity of the object of divine worship? They are open to inspection and examination, and the style of them is sufficiently clear and free from ambiguity. If they do, I retract my charge, and take shame to myself. If they do not, the charge is not the less true, because it is not acknowledged, or because both ingenious and good men may not be convinced of it. The church of Rome has had a Pascal, a Fenelon, and a Bossuet, and yet all Protestants maintain it to be an idolatrous and antichristian church; and though the church of England should be able to boast greater names than these, men who should avow and defend all her doctrines and usages, which, however, is not the case, it would not, on that account, be less idolatrous, or antichristian.

"That these articles and public offices *do* speak a language different from that of the scriptures above recited, is to me exceedingly obvious, and I cannot but think and speak according to this evidence. I shall

shall in this place recite a few passages, that others may judge whether my charge be void of all foundation or not.

"In the Nicene creed, which is adopted by the church of England, Christ is affirmed to be *God of God, light of light, very God of very God.* In the Athanasian creed, the godhead of the Son, and of the Holy Ghost, is said to be *one with that of the Father, their glory equal, their majesty coeternal.* Christ is also there said to be *uncreated, eternal, almighty and incomprehensible.* In this creed it is said, that *we are compelled by the christian verity, to acknowledge each person in the Trinity, by himself, to be God and Lord.* And, moreover, of this *catholic faith,* as it is here called, it is asserted that *except a man believe it faithfully, he cannot be saved.* The proper *articles* of the church of England are drawn up in the same style with these two creeds, asserting, that *in the unity of the godhead there be three persons, of one substance, power, and eternity, the Father, the Son, and the Holy Ghost.*

"Agreeably to this unscriptural *doctrine,* is the *practice* of this church. In the *Litany,* or form of solemn supplication, the petitions are chiefly addressed

dressed to Christ. *O God the Son, redeemer of the world, have mercy upon us miserable sinners;* and though, in the opening of it, all the three persons are addressed, both jointly and separately, yet the principal reference is evidently kept up to Christ, through the whole, as appears by these clauses; *By the mystery of thy holy incarnation, by thy holy nativity and circumcision, by thy baptism, fasting, and temptation,* &c. After this curious passage, there is no mention of any other object of worship, and the whole concludes with the solemn and repeated invocation of the Son only. *Son of God, we beseech thee to hear us. O Lamb of God, that takest away the sins of the world, grant us thy peace. O Christ hear us. Lord have mercy upon us. Christ have mercy upon us. Lord have mercy upon us.*

"In the Communion Service is the following very strange and inconsistent address to Christ. *O Lord, the only begotten Son, Jesus Christ; O Lord God, Lamb of God, Son of the Father, that takest away the sins of the world, have mercy upon us, &c. For thou only art holy, thou only art the Lord; thou only, O Christ, with the Holy Ghost, art most high in the glory of God the Father.*

"To

"To quote no more, in the Collect for Trinity Sunday, God is said to have given us *grace, to acknowledge the glory of the eternal Trinity, and in the power of the divine Majesty, to worship the Unity.*

"Now the whole of this appears to me to be directly contrary to the plain tenor of the scriptures. If, therefore, I consider the doctrine of the scriptures to be true, this contrary doctrine cannot but appear to be false; and if the doctrine of the scriptures on this subject be of *importance*, that of the church of England must, in the same proportion, appear to be *dangerous;* and I should think it criminal in myself, or any other person (who should see this subject in the same light that I do) not to bear our testimony (in whatever manner we shall judge, from our situation and circumstances, to be the most proper and effectual,) against so gross a corruption of our holy religion, whatever human laws may enjoin to the contrary.

"The act of William and Mary, which in part declares the doctrine of the divine Unity to be *blasphemy*, only expresses the opinion of William and Mary, and of those English Lords and Commoners who, if they may be said to have had any opinion at all

all about the matter, happened to think as William and Mary did. But what is that to the solemn declaration of God himself, which asserts their opinion to be impious and blasphemous.

"Whatever respect other persons may be disposed to feel for a *parliamentary system of religion*, I own that the very idea of it appears to me to be, in the highest degree, preposterous and absurd; and that I should receive with much more respect a parliamentary system of *philosophy*, and for what appears to be a very plain and sufficient reason, *viz.* that, of the two, our law-makers probably know rather more of philosophy than divinity.............." Some persons may think, that the doctrine of a Trinity in the divine Unity, is only a metaphysical subtility, of no practical importance. This subject I have argued with the Dissenter above referred to, (Letter concerning the Dissenting Interest, p. 21, &c.) and I shall not here repeat what I have before advanced on that subject. I shall only observe, in general, that the doctrine of the pre-existence of Christ, and of a plurality of persons in the Deity, appears to me to have been one of the first great corruptions of christianity, and the natural foundation for most of the rest, as will perhaps appear in the history which I hope, in due time, to publish of those corruptions."

This pamphlet has annexed to it a long and important quotation from Dr. Hartley's Observations on Man, and a short one from the writings of Nicholas Mann, Esq. in which the most serious and affecting considerations are set forth respecting the state and circumstances of the European world, both with respect to doctrinal matters, the profligacy and corruption of manners, and the judgments and calamities that may be expected to follow in consequence. If these observations of Hartley and Mann, had any weight and authority in them at the time they were written by their respective authors, or in 1773, when extracted by Dr. Priestley, they must appear to have more *now*, after the occurrence of so many astonishing events, when the cup of iniquity is more full, and the political hemisphere appears charged with fresh storms and hurricanes ready to break forth.

About this time, the Rev. Theophilus Lindsey, Vicar of Catterick in Yorkshire, a person of distinguished piety and worth, who had been one of the petitioning clergy before mentioned, finding all hopes of of reformation in the church of England hopeless, and having been long uneasy under the burden of subscription and the imposition of trinitarian forms of worship, nobly resigned his preferments in the church with very little prospect of being elsewhere provided for. He published a valuable *Apology*, in which he assigned the reasons for his resignation, and stated powerful arguments and interesting facts in favour of the Unitarian doctrine: pointing out at the same time the unscriptural forms of worship contained in the Liturgy. Soon after he published *The Book of Common Prayer Reformed, according to the plan of the late Dr. Samuel Clarke*, London, 1774. Besides the amendments of Dr. Clarke, Mr. Lindsey, with the advice of friends, made such other alterations in the Liturgy as were judged necessary to render it unexceptionable with respect to the *object* of religious worship, &c. and proposed it "as a Liturgy to be made use of by a society of like-minded christians; amongt whom he should be happy if his own labours in the ministry of the gospel might find acceptance."

Dr. Priestley, who had contracted an acquaintance with Mr. Lindsey in Yorkshire some time before,

fore, and was sufficiently zealous in the cause of truth and piety, endeavoured to forward his views by a short tract, entitled, *A Letter to a Layman, on the subject of the Rev. Mr. Lindsey's Proposal for a Reformed English Church, upon the Plan of the late Dr. Samuel Clarke*, London, 1774.

In this piece our Author endeavours to remove some objections of his young friend to Mr. Lindsey's proposal, by representing the vast importance of Christianity; the corruptions that have been introduced into it by Popery, many of which are still retained in the church of England; the necessity of separating from a corrupt church and relinquishing an antichristian worship. This, enforced by the example of the Reformers, who acted up to the light they enjoyed. False and frivolous pleas stated and confuted. The improbability of any reformation in the church of England, from the fate of the clerical and dissenting petitions. To pay a regard to splendour, rank, and external circumstances in the choice of any form of religion, or continuance in the profession of it inconsistent with the spirit of christianity. Truth to be preferred for its own sake. The uncommon merit of those who, though in high stations, act according

according to the dictates of their consciences. Mr. Lindsey's proposal of *a Reformed church of England*, with a *liturgy*, coincides with the opinion of those who object to the mode of worship prevailing among the Dissenters; consequently they can have no good reason to decline supporting him in his laudable and honourable attempt. These and other similar topics are insisted and enlarged upon with much spirit and propriety in this excellent tract, which still deserves to be read as an *incentive* to the consistent and truly christian conduct recommended in it. It is a satisfaction to think that Mr. Lindsey's sincere and honourable endeavours in religion, were crowned with a considerable degree of success; that a respectable society at Essex-street chapel, London, was formed under his care; that this society has flourished for more than thirty years, and still continues to exist under the pastoral care of the Rev. Thomas Belsham.

While Dr. Priestley resided with Lord Shelburne, he published the third volume of his Institutes of Natural and Revealed Religion, an account of which has been before given. In the preface of that work, he opposed the notions of Doctors Reid, Oswald and Beattie respecting the doctrine of *common*

mon

mon sense. Pursuing the same subject, he afterwards printed, in a separate work, an *Examination* of what each of these writers had advanced with respect to that point. During the same period, he also published Hartley's Theory of the Human Mind, on the principle of the Association of Ideas, &c. Disquisitions relating to Matter and Spirit, &c. The Doctrine of Philosophical Necessity illustrated. A Free Discussion of the Doctrines of Materialism and Philosophical Necessity, in a correspondence between Dr. Price and Dr. Priestley, &c. A Defence of the Doctrine of Necessity, in two letters, to the Rev. Mr. John Palmer. A Letter to Jacob Bryant, Esq. in defence of Philosophical Necessity. He re-printed Collins's Inquiry concerning Human Liberty, with a Preface. I barely mention these pieces in the order of time, but forbear to enlarge upon them, as they have already been considered by a gentleman of genius and ability in the course of this work: and, generally speaking, they do not enter into my plan, which is confined to matters purely theological. A part of the *Disquisitions relating to Matter and Spirit*, may, however, be considered, as having a relation to theology. In this view the Author considers

it,

it, when he refers to it, page xix, in his Preface to *The History of the Corruptions of Christianity.* "The whole of what I have called the *Sequel to the Disquisitions, (or the history of the philosophical doctrine concerning the origin of the soul, and the nature of matter, with its influence on christianity, especially with respect to the doctrine of the pre-existence of Christ)* I wish to have considered as coming properly within the plan of this work, and essential to the principal object of it. Indeed, when I published the *Disquisitions,* I hesitated whether I should publish that part then, or reserve it for this *History.* But the rest of this work was not then ready, and it was of too much use for the purpose of the other, not to go along with it. I wish the general arguments against the pre-existence of Christ, contained in *sect.* vi. of that Sequel, to be particularly attended to." There are also a few passages in the *Illustrations of the doctrine of Philosphical Necessity,* and some in the *Correspondence* with Dr. Price, which may be referred to the subject of Theology.

In the year 1777, our Author published at London, in 4to. *A Harmony of the Evangelists, in Greek; with Critical Dissertations, in English.* This work is respectfully dedicated to Dr. Price, with a preface, in which after observing that the history of Christ is infinitely more important than that of any other man that ever lived on the face of the earth, in comparison with whom, kings, law-givers, or philosophers, appear as nothing, and therefore deserves to be viewed in every possible light, he considers what has been done by former writers who have engaged in the task of harmonizing the Evangelists before him. He blames Osiander among the more ancient harmonists, and Dr. Macknight among the modern, for going upon the supposition that all the Evangelists relate every thing in chronological order, so that little or nothing is to be transposed in any of them: he on the contrary thinks, that the foundation of this hypothesis proceeds upon such a notion of the *inspiration* of the gospels, and other books of scripture, as appears to be equally indefensible and unnecessary; and that the endeavours of the friends of revelation to demonstrate the perfect harmony of the historical books of scripture, and to remove every minute contradiction in them, have not only been unsuccessful, and have thereby given the enemies of revelation a manifest advantage; but that,

that, even if they could have succeeded to their wish, the result would, in reality, have been unfavourable to the proper defence of revelation, with those who duly consider the nature of historical evidence.

He is far from thinking, however, that there is any uncertainty or ambiguity with respect to the main and important facts recorded by the Evangelists, on which our christian faith depends. p. ix. " No two persons ever gave exactly the same account of any considerable transaction, though they had the same opportunity of being well informed concerning it. On this account, differences in the narration of lesser circumstances seem to be as necessary to *complete and satisfactory evidence*, as an agreement with respect to what is capital and essential to any story. Nay, in many cases, the more persons differ in their accounts of some things, the more conclusive and satisfactory is their evidence with respect to those things in which they agree.

" It appears to me, that the history of the Evangelists has this complete evidence. They agree in their account of every circumstance of importance, which shews that their histories were written by men who were either themselves witnesses of the transactions they record, or were well informed concerning them by those who were witnesses; and yet their style, and manner of writing, their more full or more concise account of discourses, together with their very different arrangement of the parts of their narrative, and their disagreement with respect to facts of small consequence, demonstrate, in my opinion, that (excepting John, who is well known to have written some time after the rest of the evangelists) they had no communication with one another, and therefore that they are to be considered as original and independent witnesses of the same fact."

Our Author acknowledges, that he was first led into the scheme of that harmony he has adopted, by reading *Mr. Mann's Dissertations on the times of the birth and death of Christ;* and though he departed from his disposition of many particular events, yet a variety of additional arguments occurred to him in support of his general hypothesis. The method which our Author pursued in arranging the parts of his Harmony is curious, and deserves to be recorded. "I procured two printed copies of the gospels, and having cancelled one side of every sheet, I cut out all the separate histories, &c. in each gospel; and having

ing a large table appropriated to that use, I placed all the corresponding parts opposite to each other, and in such an order as the comparison of them (which, when they were brought so near together, was exceedingly easy) directed.

"In this loose order, the whole Harmony lay before me a considerable time, in which I kept reviewing it at my leisure, and changing the places of the several parts, till I was as well satisfied with the arrangement of them as the nature of the case would admit. I then fixed the places of all these separate papers, by pasting them, in the order in which they lay before me, upon different pieces of pasteboard, carefully numbered, and by this means, also, divided into sections."

The Critical Dissertations that follow contain, Observations on the time of the birth of Christ. On the time of the death of Christ. On Daniel's prophecy of seventy weeks. Mr. Lauchlan Taylor's Observations concerning the length of the reign of Xerxes. Additional arguments in support of the opinion that Xerxes reigned only eleven years, and not twenty-one. On the duration of Christ's ministry. Remarks on some of the arguments of Mr. *Mann*,

Mann, with observations in confirmation of them. Additional arguments in support of the hypothesis that Christ preached only one year and a few months. Objections to the preceding hypothesis considered. The order of the principal events in the gospel history, &c. and in general they treat of all the remarkable facts and circumstances recorded in the Evangelists. A Jewish and Julian Calendar follows for the time of the public ministry of Christ.

The Harmony itself is in Greek as the title expresses, and by being so is particularly adapted for the perusal of scholars; the original terms and phraseology of the Evangelists expressed in the language in which they wrote, and judiciously brought together under one point of view; being better fitted to convey their genuine meaning, than any translation can possibly be. The Author has printed in a *larger character* what appeared to him the most authentic, and the most circumstantial account of every important incident, collected from all the gospels promiscuouly, placing the parallel accounts in separate columns, printed in a *smaller character*. By this means, any person who would chuse to read *the whole history*, without interruption, may confine

himself

himself to the larger character, having recourse to the columns, printed in the smaller character only when he has occasion to compare the different accounts of the same thing.

Soon after the riot that took place in London, on account of the act made in favour of the Roman Catholics, our Author published, without his name, a small piece, entitled, *A Free Address to those who have petitioned for the repeal of the late act of Parliament in favour of the Roman Catholics*, London, 1780 Price twopence.

The intention of this piece was to enlighten the minds and moderate the zeal of those mistaken Protestants, who were at that time actively engaged in measures against the Roman Catholics. Our Author shews from the example and precepts of Christ, that no hostile or coercive methods ought to be used in defence of his religion, that all attempts of the kind have proved abortive; that every species of persecution, or restraint upon the consciences of men, is contrary to the spirit and genius of christianity; that the indulgence granted to the Papists, by the late act, is what humanity and sound policy loudly called for; that they are entitled to much greater liber-

ty;

ty; and that from the smallness of their numbers, and the change that there is good ground to believe has taken place in their sentiments, there is no reason to apprehend any danger from them. These, and other topics relating to the subject, are stated and urged with great force and propriety.

In 1780, Dr. Priestley published *A Harmony of the Evangelists in English; with critical Dissertations, an occasional Paraphrase, and Notes for the use of the unlearned,* 4to. London.

This Harmony is arranged in the same manner in English, as the former one was in Greek. The Critical Dissertations are also the same. The English translation is corrected throughout, wherever the Author thought it necessary. Useful notes are added to this Harmony on passages that required illustration, generally collected or supplied by the Author himself. Some were communicated by friends. Those signed T and J. were composed by the late Mr. Turner of Wakefield, and Dr. Jebb. A valuable occasional paraphrase is given, some parts of which are very fine, particularly that on the Lord's Prayer, Matth. vi. 9, &c. and on John xvii. throughout.

The

The correspondence that took place between Dr. Priestley and Dr. Newcomb, Bishop of Waterford, on the Duration of our Saviour's Ministry, may be considered as connected with the subject of these Harmonies. It took its rise from Dr. Newcomb's having, in his own Harmony, undertaken the defence of the common hypothesis of the duration of our Lord's ministry for three years or more, and having objected to what Dr. Priestley had advanced on the subject before. Two letters were published at Birmingham in 1780, addressed to the Bishop of Waterford, with respect to this point, by Dr. Priestley, the first of which had been before printed in his English Harmony, and replied to by the Bishop. Dr. Newcomb also replied to the second letter with such ingenuity and candour, as struck Dr. Priestley with admiration. This occasioned a third letter to the Bishop, on the part of our Author. Birmingham, 1781. To this last letter, Dr. Newcomb made no public reply: but wrote a private letter to Dr. Priestley, part of which the Doctor published, with the Author's consent, expressing at the same time his esteem for the Bishop, and the amicable manner in which the controversy had been conducted.

Dr.

Dr. Priestley, while he remained with Lord Shelburne, accompanied that nobleman in an excursion to the Continent, and having had occasion to converse frequently with unbelievers, and hear their sentiments, conceived he should be able to combat their prejudices with advantage, and provide some antidote against the baneful progress of infidelity. With this view, he composed and published the first part of his *Letters to a Philosophical Unbeliever*, bearing date *Calne*, March 1780. The vast importance of the subjects treated of in this publication, are very justly stated by the Author in the opening of a very valuable preface. "It will I think be acknowledged by all persons who are capable of reflection, and who *do* reflect, that, in the whole compass of speculation, there are no questions more interesting to all men than those which are the subject of these *letters*, viz. Whether the world we inhabit, and ourselves who inhabit it, had an intelligent and benevolent author, or no proper author at all? Whether our conduct be inspected, and we are under a righteous government, or under no government at all? And, lastly, whether we have something to hope and fear beyond the grave, or are at liberty to adopt the

Epicurean maxim, *Let us eat and drink, for to-morrow we die.*" The first letter treats of the nature of evidence in general with respect to subjects that are capable of strict demonstration, and others which though they are not, yet admit of a sufficiently satisfactory evidence. He observes, "It is not pretended, that the evidence of the propositions in natural or revealed religion, is always of the former of these two kinds, but generally of the latter, or that which depends on the association of ideas; and in revealed religion, the evidence chiefly arises from testimony, but such testimony as has never yet been found to deceive us. I do not therefore say, that I can properly *demonstrate* all the principles of either; but I presume that, if any person's mind be truly unprejudiced, I shall be able to lay before him such evidence of both, as will determine his assent; and, in some of the cases, his persuasion shall hardly be distinguishable, with respect to its *strength*, from that which arises from a demonstration properly so called, the difference being, as mathematicians say, less than any assignable quantity."

The second letter contains the direct evidence for the being of a God. Our Author reasons from effects to causes. Men make *chairs* and *tables*, build *houses* and write *books*, and chairs, tables, houses or books, are not made without men. Birds build nests, spiders make webs, bees make honeycombs, &c. One plant proceeds from another, and one animal from another, by natural vegetation, or generation, and therefore it is concluded that every plant and every animal had its proper parents. Reasoning upon these and other similar facts that constant experience affords to human observation, our Author lays it down as a rule that is universally true, that *Nothing begins to exist without a cause.* If a *table* or *chair* must have had a designing cause, capable of comprehending their nature and uses, the *wood*, or the *tree*, of which the table was made, and also the *man* that constructed it, must likewise have had a designing cause, &c. For the same reason that the human species must have had a designing cause, all the species of brute animals, and the *world* to which they belong, and with which they make but *one system*, and indeed all the visible *universe*, (which, as far as we can judge, bears all the marks of being *one work*) must have had a cause or author, possessed of what we may justly call *infinite power* and intelli-gence.

gence. It follows, therefore, from the most irresistible evidence, that the world must have had a designing cause, distinct from, and superior to itself. This conclusion follows from the strongest analogies possible. It rest on our own constant experience; and we may just as well say, that a *table* had not a designing cause, or no cause distinct from itself, as that the *world*, or the *universe*, considered as one system, had none. This necessary cause we call *God*, whatever other attributes he be possessed of.

In the third Letter, various atheistical schemes and suppositions are considered and confuted. In the fourth Letter, an account is given of the necessary attributes of the original cause of all things, in which it is demonstrated, First, that this Being must be what we term *infinite*, or without limitation in knowledge and power. Secondly, that he must be *omnipotent* or occupy all space, though this attribute is equally incomprehensible by us with the infinite extent of his power and knowledge. Thirdly, that this infinite Being, who has existed without change, must continue to exist without change to eternity, is likewise a conclusion that we cannot help drawing, though the subject being incomprehensible, we may

not

not be able to complete the demonstration. "Fourthly, There cannot be more than *one* such Being as this. Though this proposition may not be strictly demonstrable by us, it is a supposition more natural than any other, and it perfectly harmonizes with what has been strictly proved and deduced already. Nay, there seems to be something hardly distinguishable from a contradiction in the supposition of there being *two infinite Beings of the same kind*, since, in idea, they would perfectly *coincide*. We clearly perceive, that there cannot be two *infinite spaces*, and since the analogy between this infinite unintelligible Being, as we may call it, and the infinite intelligent one, has been seen to be pretty remarkable in one instance, it may be equally strict here; so that, were our faculties equal to the subject, and had we proper *data*, I think we should expect to perceive, that there could no more be two infinite, intelligent, and omnipresent Beings, than there can be two infinite spaces.

"Indeed, their being *numerically two*, would in in some measure limit one another, so that, by the reasoning we have hitherto followed, neither of them could be the originally existent Being. Supposing them to be equally omnipotent, and that one of them should

should intend to do, and the other to undo, the same thing, their power would be equally balanced; and if their intentions always coincided, and they equally filled all space, they would be as much, and to all intents and purposes, *one and the same Being*, as the coincidence of two infinite spaces would make but one infinite space.

The fifth Letter, contains the evidence for the general benevolence of the Deity. The sixth Letter proposes arguments for its infinite extent. The seventh Letter exhibits the evidence of the moral government of the world, and the branches of natural religion. The eighth Letter treats of the evidence for the future existence of man. In the ninth Letter, the strange and ridiculous paradoxes of Mr. Hume, in his Dialogues on Natural Religion, are examined and exposed. The tenth Letter contains an Examination of Mr. Hume's Essay on a particular Providence, and a Future State. In the eleventh letter, the sceptical and atheistical reasonings contained in a French publication, entitled the *Systeme de la Nature*, are considered. The twelfth Letter contains an Examination of some fallacious methods of demonstrating the being and attributes of God, in which our Author differs

differs from the celebrated Dr. Clarke in some particulars. The thirteenth Letter treats of the ideas of *Cause* and *Effect*, and the influence of Mr. *Hume's* opinion on this subject in the argument for the being of a God. The fourteenth Letter contains an Examination of Mr. Hume's metaphysical writings, in which our Author appears to entertain but a low idea of him as a metaphysical and moral writer, detects his fallacious reasonings, and asserts that he had no idea of the power of association in the human mind, &c.

In 1782, our Author published at Birmingham seven *Additional Letters to a Philosophical Unbeliever*, occasioned by a publication in favour of atheism, by a person who called himself William Hammon, jun. and avowed himself an atheist. In these letters, the arguments and reasonings of Mr. Hammon are considered and replied to.

In 1787, Dr. Priestley completed his plan, by publishing at Birmingham, *Letters to a Philosophical Unbeliever, Part II. containing a State of the Evidence of revealed Religion, with Animadversions on the two last chapters of the first volume of Mr. Gibbon's History of the Decline and Fall of the Roman Empire.*

These

These Letters are sixteen in number. The five first treat of the nature of testimony, the evidence of Revelation, its antecedent probability, the nature of prejudice for or against it, the causes of infidelity in persons of a speculative turn of mind. The sixth, gives the history of the Jewish religion. The seventh, the historical evidence of the truth of christianity. The eighth, assigns the causes of infidelity in early times. The ninth, gives a more particular account of the nature of those prejudices to which the heathens were subject with respect to christianity. The tenth, describes the different foundations on which the belief of Judaism or Christianity, and that of other religions stands. The eleventh, compares the evidence of Judaism and Christianity with that of Mahometanism, and of the religion of Indostan. The twelfth, treats of the nature of idolatry, and the attachment of the Heathens to it, as a principal cause of the hatred of christians. In the thirteenth, the attachment of the heathens to their religion is more particularly proved. The fourteenth, treats of the objections to the historical evidence of christianity in early times. The fifteenth, of other objections to christianity in early times. The sixteenth and last

contains,

contains, as expressed in the title, animadversions on the first volume of Mr. Gibbons's history of the decline and fall of the Roman empire.

These are the contents of this important publication, all of which are deserving of an attentive perusal. The historical evidence of the Jewish and Ch istian revelations are stated with such force and precision, that it is impossible to account for the rise and progress of either without admitting the truth of the miraculous facts recorded in the scriptures. Upon the whole, these letters, to a Philosophical Unbeliever, form a very valuable compendium of the arguments in favour of natural and revealed religion, from which all may derive utility, but are particularly calculated for the improvement and benefit of those persons who have not leisure or inclination to peruse large and voluminous treatises.

We now proceed to give an account of a celebrated work of Dr. Priestley's; a work which had been long projected by its Author, but delayed from time to time, and which gave rise to a controversy that continued for several years, *viz. An History of the Corruptions of Christianity, in two volumes*. Birmingham, 1782. This publication was originally

promised

promised on a much smaller scale, viz. as a Sequel to the Author's *Institutes of Natural and Revealed Religion*, but having extended his views considerably, he thought proper to make it a separate work. To this Treatise is prefixed an affectionate and pathetic dedication to the Author's friend, the Rev. Mr. Lindsey, and a preface giving an account of his views and intentions in the composition of it. The general division of the work is into thirteen parts, each of which is sub-divided into sections, besides some appendices and a general conclusion, viz.

Part 1. The History of Opinions relating to Jesus Christ.

2. The History of Opinions relating to the Doctrine of Atonement.

3. The History of Opinions concerning Grace, Original Sin and Predestination.

4. The History of Opinions relating to Saints and Angels.

5. The History of Opinions concerning the State of the Dead.

6. The History of Opinions relating to the Lord's Supper.

Part 7. The History of Opinions relating to Baptism.

Appendix to Parts 6 and 7, containing the History of the other Sacraments, besides Baptism and the Lord's Supper.

8. A History of the changes that have been made in the method of conducting Public Worship.

9. The History of Church Discipline.

10. The History of Ministers in the Christian Church, and especially of Bishops.

11. The History of the Papal Power,

Appendix 1 to Parts 10 and 11.

The History of Councils,

Appendix 2, to Parts 10 and 11.

Of the Authority of the Secular Powers, or the Civil Magistrate, in Matters of Religion,

Appendix 3, to Parts 10 and 11.

Of the Authority of Tradition and the Scriptures, &c.

12. The History of the Monastic Life:

13. The History of Church Revenues.

The General Conclusion, containing,

Part 1. Considerations addressed to Unbelievers,

believers, and especially to Mr. Gibbon.

Part 2. Considerations addressed to the advocates for the present establishments of Christianity, and especially Bishop Hurd.

Appendix, containing a summary view of the evidence for the primitive christians holding the doctrine of the simple humanity of Christ.

Many curious facts and particulars are recorded under each of the parts above-mentioned; and the progressive changes, and successive stages of corruption, are marked out and delineated in the sub-divisions or sections; so that this work may be considered as an ecclesiastical history, composed upon a new plan, and exhibited under a peculiar form. The history of opinions relating to Jesus Christ, and that of the doctrine of atonement, occupy however by far the largest space, comprehending the greater part of the first volume.

With respect to the doctrine of atonement, I find nothing materially new added to what the Author had before advanced in his Treatise on *the one great end of the life and death of Christ*, (of which a copious account

account has already been given) until page 213, where the proper history of the doctrine commences. The Author contented himself, as he mentions in his preface, with giving the substance of his former work on the subject, which he has done very ingeniously and agreeably. The historical part, however, of this work is entirely new, and comprehends an account of the opinions of the apostolical fathers, of the fathers till after the time of Austin, of the state of opinions from the time of Austin to the reformation, and of the doctrine of the reformers on this subject. In treating of the opinion of the apostolical fathers, our Author observes, p. 214, "It cannot be determined from the primitive christians calling the death of Christ *a sacrifice for sin*, a *ransom*, &c. or from their saying, in a general way, that Christ died in our stead, and that he *bore our sins*, or even if they carried this figurative language a little farther, that they really held what is now called *the doctrine of atonement*, viz. that it would have been inconsistent with the maxims of God's moral government to pardon any sin whatever, unless Christ had died to make satisfaction to divine justice for it. Because the language abovementioned may be made use of by

persons

persons who only believe that the death of Christ was a necessary circumstance in the scheme of the gospel, and that this scheme was necessary to reform the world." And after quoting several passages from Clemens Romanus, Barnabas, and the Shepherd of Hermas, he adds, "It seems pretty evident, that *so far* we find no real change of opinion with respect to the efficacy of the death of Christ. These writers adopt the language of the apostles, using the term *sacrifice* in a figurative sense, and representing the value of good works, without the least hint or caution, lest we should thereby detract from the merits of Christ, and the doctrine of salvation by his imputed righteousness."

Various quotations are introduced from Cyprian, Origen, Athanasius, Lactantius, Gregory Nazianzen, Austin, &c. concerning the import of which the Author has the following remark: P. 246. "Upon the whole, I think it must appear sufficiently evident, that the proper doctrine of atonement was far from being settled in the third or fourth centuries, though some little approach was made towards it, in consequence of supposing that what is called a *ransom* in a figurative sense, in the New Testament, was something

thing more than a figure; and therefore that the death of Christ was truly a *price paid for our redemption*, not indeed directly from *sin*, but rather from *death*, though it was not settled *to whom* this price was paid. In general, the writers of those times rather seem to have considered God as the person who paid the price, than he that received it. For, man being delivered into the power of the devil, they considered the price of redemption as paid to him. As to the forgiveness of sins, it was represented by all the Fathers, and even by Austin himself, as proceeding from the free grace of God, from which free grace he was farther induced to give up his son, as the price of our redemption from the power of the devil. We must therefore proceed farther, before we come to any regular system of atonement, founded on fixed principles, such as are now alleged in support of it."

Our Author proceeds to quote and give the opinions of Gregory the Great, Peter Lombard, Thomas Aquinas, and other writers, till the period of the reformation, when by the labours of Wickliffe, Luther, Calvin, and others, whose writings he quotes, the doctrine began to assume the appearance of a system, though not without some diversity of opinion even amongst the orthodox themselves. Faustus

Socinus

Socinus and Crellius are mentioned, as bearing their testimony against the prevailing doctrine, and the whole is concluded with a train of reflections arising from the subject.

We now turn to the first part of the work: *The history of opinions relating to Jesus Christ.* This part is divided into eleven sections. After shewing in the introduction, that the unity of God and humanity of Christ are the clear doctrines of the scriptures, the Author proceeds to collect evidence for the last of these facts from ecclesiastical antiquity. In the first section, he inquires into the opinion of the ancient Jewish and Gentile churches, and alleges the testimonies of Epiphanius, Origen, and Eusebius, to prove that the Ebionites and Nazarenes, by which names the Jewish christians were distinguished, held the humanity of Christ; some believing his miraculous conception, and others not. He also quotes a very striking passage from Athanasius to the same effect, viz. that "all the Jews were so firmly persuaded, that their Messiah was to be nothing more than a man like themselves, that the apostles were obliged to use great caution in divulging the doctrine of the proper divinity of Christ." Here Dr. Priestley

Priestley very properly remarks, "But what the apostles did not teach, I think we should be cautious how we believe. The apostles were never backward to combat other Jewish prejudices, and certainly would have opposed this opinion of theirs, if it had been an error. For if it had been an error at all, it must be allowed to have been an error of the greatest consequence."

Our Author observes, p. 14, "Of the same opinion with the Nazarenes or Ebionites among the Jews, were those among the Gentiles whom Epiphanius called *Alogi*, from their not receiving, as he says, the account that John gives of the *Logos*, and the writings of that apostle in general. But Lardner, with great probability, supposes * there never was any such heresy as that of the *Alogi*, or rather that those to whom Epiphanius gave that name, were unjustly charged by him with rejecting the writings of the apostle John, since no other person before him makes any mention of such a thing, and he produces nothing but mere hearsay in support of it. It is very possible, however, that he might give such an ac-

* History of Heretics, p. 447.

count

count of them, in consequence of their explaining the *Logos* in the introduction of John's gospel in a manner different from him and others, who in that age had appropriated to themselves the name of orthodox.

Dr. Priestley also produces a very full testimony from Justin Martyn, in favour of the existence of Unitarian christians and believers in the proper humanity of Christ in his time; and in p. 18 refers to Eusebius, as relating "that the Unitarians in the primitive church, always pretended to be the oldest christians, that the apostles themselves had taught their doctrine, and that it generally prevailed till the time of Zephyrinus, bishop of Rome; but from that time it was corrupted." He also thinks that the apostle John meant to approve the doctrine of those who held that Christ was truly a man, when he says, 1 Ep. iv. 3. *Every spirit that confesses that Jesus Christ is come in the flesh, is of God;* and that he intended to censure the opinion of the *Docetæ*, or those who denied the reality of our Lord's humanity, by saying, *every spirit which confesses not that Jesus Christ is come in the flesh, is not of God, and this is that spirit of antichrist, whereof ye have heard that it should*

 come,

come, and even now already is it in the world." For this was the first corruption of the christian religion by the maxims of heathen philosophy, and which proceeded afterwards, till christianity was brought to a state little better than paganism." He also lays some stress on the circumstance "that Hegesippus, in giving an account of the heresies of his time, not only makes no mention of this supposed heresy of the Nazarenes or Ebionites, but says that in his travels to Rome, where he spent some time with Anicetus, and visited the bishops of other sees, he found that they all held the same doctrine, that was taught in the law, by the prophets, and by our Lord. What could this be but the proper Unitarian doctrine, held by the Jews, and which he himself had been taught."

Our Author concludes this section in the following words: "It is remarkable that as the children of Israel retained the worship of the one true God all the time of Joshua, and of those of his cotemporaries who outlived him, so the generality of Christians retained the same faith, believing the strict unity of God, and the proper humanity of Christ, all the time of the apostles, and of those who conversed with them, but began to depart from that doctrine present-

ly

ly afterwards; and the defection advanced so fast, that in about one century more, the original doctrine was generally reprobated, and deemed heretical."

The second section treats of the first step that was made towards the deification of Christ, by the personification of the Logos. This our Author ascribes to the operation of several causes. The disgust that was taken by many, and particularly by philosophers, at the doctrine of a crucified Saviour, concerning which there are plain traces to be found in scripture. The allegorical method of interpreting scripture adopted by some learned Jews, particularly Philo, and imitated by Christians. The oriental doctrine of emanations from the great original mind, and that all spirits whether dæmons, or the souls of men, were of this divine origin. The prevalence of the doctrine of Plato, who styled the Logos a second God, according to Lactantius. A mistaken apprehension of the meaning of John in the beginning of his gospel, and supposing that the Logos there mentioned signifies the person of Christ, and not an attribute of God himself. Full of these erroneous notions, the fathers of the second and third centuries, several of whom had been converts from Paganism,

and

and Platonic philosophers before their conversion, particularly Justin Martyr, soon corrupted the simple doctrine of the gospel, as delivered by the apostles, and introduced a second God into their system of christianity. Passages from Justin Martyn, Theophilus, Tatian, Athenagoras, Tertullian, Novatian, &c. are quoted in proof of this assertion, and as a specimen of their method of interpreting scripture.

The third section. That supremacy was always ascribed to the Father before the council of Nice, is proved clearly by quotations from various fathers of the second and third centuries, and some beyond that period. Yea, even the fathers of the council of Nice themselves, by calling Christ *God of God*, could not mean that he was strictly speaking equal to God the Father.

The fourth section treats of the difficulty with which the doctrine of the divinity of Christ was established. It is here shewn how extremely averse the more numerous and unlearned part of christians were to receiving the doctrine of the Trinity and the divinity of Christ even in the most qualified form, and to what pains and shifts the philosophising part of the clergy were driven to make even a tolerable

defence

defence of their opinions. The following quotations, among many others, are worthy of particular note: "The simple, the ignorant, and the unlearned, "(says Tertullian) who are always a great part of the "body of christians, since the rule of faith itself," (meaning perhaps the *apostle's creed*, or as much of it as was in use in his time) "transfers their worship "of many Gods to the one true God, not under-"standing that the unity of God is to be maintained, "but with the *œconomy*, dread this *œconomy*, imagin-"ing that this number and disposition of a trinity is "a division of the unity. They therefore will have "it, that we are worshippers of two, and even of three "Gods; but that they are the worshippers of one "God only. We, they say, hold the *monarchy*. "Even the Latins have learned to bawl out for mo-"narchy, and the Greeks themselves will not under-"stand the œconomy."

Origen says, "that to the carnal they taught the "gospel in a literal way, preaching Jesus Christ and "him crucified, but to persons farther advanced, and "burning with love for divine celestial wisdom" (by which he must mean the philosophical part of their audience) "they communicated the Logos."

Epiphanius

Epiphanius says, that when a Sabellian met the orthodox, they would say, " My friends, do we be-" lieve one God or three ?"

Basil complains of the popularity of the followers of Marcellus, whose disciple Photinus is said to have been, at the same time that the name of Arius was execrated. " Unto this very time," says he, in his letter to Athanasius, " in all their letters " they fail not to anathematize the hated name of " Arius ; but with Marcellus, who has prophanely " taken away the very existence of the divinity of " the only begotten Son, and abused the significa-" tion of the word *Logos*, with this man they seem " to find no fault at all."

These quotations, and others in this section, make it abundantly evident, that the doctrines of the divine Unity, and the proper humanity of Christ, had taken deep root in the minds of the generality of christians, and what can this be ascribed to, but that these doctrines had been conveyed down to them in succession from the apostles themselves.

The fifth section gives an account of the Unitarians before the council of Nice. Our Author observes, " that the Christian church in general held this

this doctrine until the time of Victor, was the constant assertion of those who professed it about this time, and I think I have shewn that this was true. He mentions several men of learning who continued to profess this doctrine afterwards, viz. Theodotus of Byzantium, Artemon, Praxeus the Montanist, Noætus, Sabellius, Paul bishop of Samosata, Beryllus of Bostra, and Photinus bishop of Sirmium. The remaining six sections of this part, treat of the Arian controversy. The doctrine concerning the Holy Spirit. The history of the doctrine of the Trinity from the councils of Nice and Constantinople, until after the Eutychian controversy. The state of the doctrine of the Trinity in the Latin church. The history of the doctrine of the Trinity after the Eutychian controversy. A general view of the recovery of the genuine doctrine of christianity concerning the nature of Christ.

Five of the sections, of which the contents have been here given, represent little else but the constant progress of error, hardening and confirming itself by degrees, and framing at last a stupendous fabric of contradiction and absurdity, guarded by penal statutes and imperial edicts. Our Author remarks

very

very justly, p. 113, "Thus, to bring the whole into a short compass, the first general council gave the Son the same nature with the Father, the second admitted the Holy Spirit into the Trinity, the third assigned to Christ a human soul in conjunction with the eternal Logos, the fourth settled the hypostatical union of the divine and human nature of Christ, and the fifth affirmed, that in consequence of this union the two natures constituted only one person."

The eleventh and last section gives an account of the revisal of the genuine doctrine concerning Christ at the Reformation, by Faustus Socinus and others, the notions of the modern Arians, and the different schemes and systems that have been adopted by some modern Trinitarian writers.

It was not to be expected that a work like the history of the corruptions of christianity, in which the Author attempts to wrest the argument from ecclesiastical antiquity out of the hands of Trinitarians and Arians, and represents the evidence arising from thence as favouring the Unitarians, should pass without animadversion. The first attack, however, came from a quarter little expected. Mr. Badcock, who (as afterwards appeared) at that time wrote in the theological

theological department of the Monthly Review, not contented as became a Reviewer, with giving a fair and candid account of the work, entered into a serious refutation of the first part, and threw out illiberal reflections on the writer. This was in June 1783. Our Author, without loss of time, composed an answer, bearing date July 21, which made its appearance in August following, entitled, *A Reply to the Animadversions on the History of the Corruptions of Christianity, in the Monthly Review for June* 1783; *with additional Observations relating to the doctrine of the Primitive Church concerning the person of Christ.* Birmingham, 1783. In this pamphlet, after some observations on the unfair and uncandid conduct of the Reviewer towards him, and proposing some emendations to his history of the corruptions, &c. he treats, section 1. Of the Nazarenes, Ebionites, and Alogi. 2. Of the inferences from Hegesippus. 3. Of what may be inferred from Justin Martyr concerning the state of opinions in his time. 4. Of the quotation from Eusebius; and Tertullian's account of the ancient Unitarians, more particularly considered, 5. Of his being charged with advancing that Justin Martyr was the first who started the notion of Christ's pre-existence. 6. Of the doctrine of the miraculous conception. 7. Of Miscellaneous Articles, in which he acknowledges one or two mistakes, but of no consequence to the main argument. In these different sections, he meets the objections of the Reviewer, and confirms what he had before advanced.

Mr. Badcock did not stop at his first attack; but in the Monthly Review for September, he attempted an elaborate refutation of Dr. Priestley's reply, and laid aside the character of a Reviewer completely to assume that of a controversial writer. Our Author has some remarks on that article of the Monthly Review, in his letters to Dr. Horsley, p. 148, &c. and promises a more particular reply on certain conditions, p. 137: "To shew that I do not say this merely to get rid of the business, I declare, that if any person, *giving his name*, shall request my attention to any particular part of it, and procure me a place in the *Monthly Review*, I will speak to it as fully and explicitly as I can; and I do not think that I should require much room to give a very satisfactory answer to any article in it. I only wish for a public and impartial hearing. In the name of truth, I only

say

say δὸς που ϛω." This intimation was not attended to, however fairly proposed. And though the denial of a place in the Monthly Review was unjust with respect to Dr. Priestley, yet it was immaterial with respect to the argument; for all the main points in controversy are discussed in the correspondence that followed between him and Dr. Horsley.

This controversial correspondence took its rise from a charge delivered by Dr. Horsley to the clergy of the archdeaconry of St. Albans, at a visitation holden May 22d, 1783, and afterwards published at London (with additions), at the request of the clergy. In this charge, Dr. Horsley entertains his clergy with remarks on the first part of the History of the Corruptions of christianity, which he affects to treat as a very superficial and contemptible performance, abounding with misrepresentations, mistakes, and inaccuracies. He accuses Dr. Priestley of reviving the arguments of Zuicker and Episcopius, which had been long ago confuted by Bishop Bull, without attempting to make them good against the objections of a writer of Dr. Bull's eminence. Besides this, he pretends to give nine specimens of insufficient proof contained in Dr. Priestley's history, the

two

two first of which, he says, are instances of the circulating syllogism. First, in alleging his own sense of scripture as a proof that the primitive faith was Unitarian, without proving the fact. Secondly, in alleging the pretended silence of St. John, about the error of the Unitarians, in proof that the Unitarian doctrine is no error, but the very truth of the gospel. Thirdly, in citing a testimony from Athanasius that does not exist, or in inferring from it that those were Jewish christians, who were only unconverted Jews. Fourthly, in making a gratuitous assumption, that the Nazarenes and the Hebrew christians were the same people, and that the faith of the Nazarenes was Unitarian. Fifthly, in misrepresenting the sense of Eusebius, and charging him with inconsistency, because another writer, who is quoted by him, speaks of Theodotus, who appeared about the year 190, as the first who held that our Saviour was a mere man, &c. Sixthly, in objecting to the doctrine of the church, from the resemblance which he finds between it and the Platonic doctrine, which resemblance rather corroborates than invalidates the traditional evidence of the Catholic faith, as, when fairly interpreted, it appears to be nothing less than the consent of the latest and earliest revelations. Seventhly, in bringing proofs of an oblique and secondary kind, that the doctrine of our Lord's divinity was an innovation of the second age, without a distinct previous proof, that the faith of the first age was Unitarian. Eighthly, a mistake in translating a passage in Athenagoras, which shews him to be a child in Platonism. Ninthly, a mistake in translating a passage of Theophilus.

These pretended specimens of insufficient proof, are aggravated and amplified with great arrogance and self-sufficiency in the course of Dr. Horsley's charge, which, though written in an elegant style, is full of rudeness and sarcastic asperity. In the Appendix, he takes notice of Dr. Priestley's reply to the Monthly Review for June, in which the same want of candour is visible as in the charge.

Dr. Priestley was not slow in vindicating his history from the attack thus made upon it. In a short time after the publication of Dr. Horsely's charge, a reply under the following title appeared....*Letters to Dr. Horsley, in answer to his Animadversions on the History of the Corruptions of Christianity. With additional Evidence that the Primitive Christian Church was Unitarian.* Birmingham, 1783. A pretty

pretty large preface is prefixed to this piece, containing remarks on the state of the controversy, the influence it had had on the mind of the public, with an account of the changes that had taken place in the Author's religious opinions. The reply consists of an introduction, eight letters, a concluding one, and a postscript. In the introductory letter, our Author says, in answer to Dr. Horsley's assertion of his arguing in a circle, "Had I produced no other proof of the Unitarianism of the *scriptures*, besides that of the *primitive church*, and also no other proof of the Unitarianism of the primitive church, besides that of the scriptures, I should have argued in a circle. But you will find that I have been far from doing this.

"Is it not usual with all writers who wish to prove *two things*, which mutually prove each other, to observe that they *do* prove each other; and therefore, that whatever evidence can be alleged for either of them, is fully in point with respect to the other? Now this is all that I have done with respect to the Unitarianism of the scriptures, and of the primitive church, which prove each other; only that, in my history, I do not profess to enter into the separate

proofs

proofs of the Unitarian doctrine from the scriptures."

In the first letter, our Author shews, in opposition to Dr. Horsley, that the Greek pronoun αυτος in the introduction to John's gospel may refer to any thing that is of the same gender in the Greek language, whether it be of a person or not. In proof of this sense of the pronoun, he quotes or refers to, various places in the New Testament. He maintains that the phrase *coming in the flesh*, as applied to Christ by John, 1 Ep. iv. 2. refers only to his being a real and true man, without any reference to a pre-existent state, and refers to other scriptural expressions as throwing light upon this phrase. He interprets a passage from Clemens Romanus differently from Dr. Horsley, and considers the epistles of Ignatius as of very doubtful authority. The second letter, treats of the distinction between the Ebionites and the Nazarenes. Here our Author quotes several passages from Epiphanius and Origen, to prove that the Ebionites and Nazarenes were agreed in sentiments with respect to the real humanity of Christ, some of which speak very plain to the point, particularly the following from Origen: "When

"you

" you consider what belief they, of the Jewish race,
" who believe in Jesus, entertain of their redeemer,
" some thinking that he took his being from Mary
" and Joseph, some indeed from Mary only and the
" Divine Spirit, but still without any belief of his di-
" vinity, you will understand," &c.

Dr. Horsley had before quoted this passage in his Appendix, and endeavoured to diminish the force of it. Our Author quotes his words, and subjoins his own remarks as follow, p. 21: " That the Jew-
" ish converts were remarkably prone to the Ebion-
" æan heresy, from which the Gentile churches in
" general were pure, is the most," you say, p. 77,
" that can be concluded from this passage, strength-
" ened as it might be with another somewhat to the
" same purpose, in the commentaries upon St. John's
" gospel. But what if it were proved that the whole
" sect of the Nazarenes was absorbed in the Ebionæ-
" an heresy in the days of Origen? What evidence
" would that afford of the identity of the Nazarenes
" and the Ebionites in earlier times? And even that
" identity, if it were proved, what evidence would it
" afford, that the church of Jerusalem had been ori-
" ginally

"ginally Unitarian under her first bishops of the cir-
"cumcision?"

"I answer, that if the Jewish christians were universally Ebionites in the time of Origen, the probability is, that they were even generally so in the time of the apostles; and that their heresy, as it is called, did exist in the time of the apostles, is abundantly evident. Whole bodies of men do not very soon change their opinions. And if, as you allow, the Jewish christians were distinguished by the name of Nazarenes (whom I think I have proved to be the same with the Ebionites, who all believed Christ to be a mere man) from the time that they were settled in the country beyond the sea of Galilee, you carry the opinions of the Ebionites, as universally held by the Jewish christians, to the very age of the apostles; for they retired into that country on the approach of the Jewish war, about which time the apostles went off the stage.

"Since all the Jewish christians were called Nazarenes or Ebionites, and all the writers that mention them speak of the doctrine of those sects *in general*, and not those *of their own time* in particular, as being that Christ was a mere man; the natural inference

 is,

is, that those sects, or the Jewish christians, did *in all times*, after they became so distinguished (which is allowed to have been just before, or presently after the destruction of Jerusalem) hold that doctrine. And supposing this to have been the case, is it not almost certain, that the apostles themselves must have taught it? Can it be supposed that the whole Jewish church should have abandoned the doctrine of the divinity of Christ, within so few years after the death of the apostles, if they had ever received it from them? As far as I yet see, Jewish christians who were not Nazarenes, or Ebionites, or Nazarenes who held any other doctrine concerning Christ than that he was a mere man, are unknown in history, and have no existence but in imagination."

In the third letter, our Author shews, that the primitive Unitarians were not considered as heretics. In opposition to Dr. Horsley, who denies the fact as asserted by Episcopius, he shews that this was not only the opinion of that writer, but also of Mosheim, who says, vol. i. p. 191, "However ready many may "have been to embrace this erroneous doctrine, it "does not appear that this sect formed to themselves "a separate place of worship, or removed themselves "from

"from the ordinary assemblies of christians." "But does it not also follow from the same fact, that these Unitarians were not expelled from christian societies by others, as they certainly would have been, if they had been considered as heretics? He shews by a quotation from the same Mosheim, that the Gnostics were in a different situation, and held separate assemblies from the church. He quotes several fathers to prove that heretics were in a state of separation from the church. He shews that Tertullian did not consider Unitarians as excluded from the name and assemblies of christians from what he says concerning the *apostles creed*, as the only proper standard of faith; for no article in that creed censures the opinions of the Unitarians but only those of the *Gnostics*, and it might have been subscribed in the time of Tertullian by any Unitarian who believed the miraculous conception." The Ebionites, being Jews, had little communication with the Gentiles, and therefore, of course, held separate assemblies; but the Alogi, who held the same doctrine among the Gentiles, had no separate assemblies, but worshipped along with other christians."

Our

Our Author observes, p. 33, "There is no instance, I believe, of any person having been excommunicated for being an Unitarian before Theodotus, by Victor bishop of Rome, the same that excommunicated all the eastern churches, because they would not celebrate Easter on the day that he prescribed. Whereas had the universal church been Unitarian from the beginning, would not the first Unitarians, the first broachers of a doctrine so exceedingly offensive to them, as in all ages it has ever been, have experienced their utmost indignation, and have been expelled from all christian societies with horror.

"What makes it more particularly evident, that the doctrine of the simple humanity of Christ was not thought deserving of excommunication in early times, is, that though the Ebionites were *anathematized*, as Jerom says; or excommunicated, it was not on account of their denying the doctrine of the divinity of Christ, but *only* on account of their rigid observance of the Mosaic law." Our Author takes notice of the alarm that the Trinitarian doctrine gave to Unitarian christians as it began to unfold itself, expressed by Tertullian by the strong words

expavescere

expavescere and *scandalizare*, and by Origen in words equally strong, as ταρασσειν, &c. From these and other circumstances, he concludes that the Unitarians must have been the majority among christians, and that the fact of their remaining in the church, and not being expelled from it, cannot be explained otherwise.

The fourth letter treats of the inference that may be drawn from the passage of Athanasius, concerning the opinion of the early Jewish christians relating to Christ. Here he gives the passage at greater length, vindicates his former interpretation of it against the exceptions of Dr. Horsley, and shews that the believing Jews and christian Gentiles are chiefly, if not altogether intended in it, and that Beausobre and the Latin translator of Athanasius, both Trinitarians, had the same ideas of the passage as himself. He enters largely into the consideration of the passage, and infers from the general tenor and connections of it, that " it can hardly be doubted but that Athanasius himself must have considered the christian church in general as Unitarian, in the time of the apostles, at least till near the time of their dispersion and death." The following observation expressed

pressed in a note, p. 47, has a great deal of force in it. "According to Athanasius, the Jews were to be well grounded in the belief of Jesus being the Christ, before they could be taught the doctrine of his divinity. Now if we look into the book of Acts, we shall clearly see that they had not got beyond the first lesson in the apostolic age; the great burden of the preaching of the apostles being to persuade the Jews that *Jesus was the Christ.* That he was likewise *God,* they evidently left to their successors; who, indeed, did it most effectually, though it required a long course of time to do it." In corroboration of his argument, our Author produces some passages from Chrysostom, in which that Father ascribes the same cautious procedure to the apostles in divulging the divinity of Christ, that Athanasius had done before him.

Our Author justly remarks, p. 52, "I cannot help observing how extremely improbable is this account of the conduct of the apostles, given by Athanasius, Chrysostom, and other orthodox fathers of the church, considering what we know of the character and the instructions of the apostles. They were plain men, and little qualified to act the cau-

tious

tious part here ascribed to them. And their instructions were certainly to teach all that they knew, even what their master communicated to them in the greatest privacy. Whereas, upon this scheme, they must have suffered numbers to die in ignorance of the most important truth in the gospel, lest, by divulging it too soon, the conversion of others should have been prevented. The case evidently was, that these fathers did not know how to account for the great prevalence of the Unitarian doctrine, among the Gentiles as well as the Jews, in the early ages of christianity, but upon such an hypothesis as this.....Let their successors do better if they can."

The fifth letter contains an argument for the late origin of the doctrine of the divinity of Christ, from the difficulty of tracing the time in which it was first divulged. Here our Author very properly requests Dr. Horsley's opinion with respect to the *time* when this great secret of Christ not being merely a man, but the eternal God himself, or the maker of heaven and earth under God, was communicated, first to the apostles themselves, and then by them to the body of christians. With this view he proposes several periods in the evangelical history, and the acts of the apostles,

apostles, without being able to find any such discovery. "To answer the charge of holding *two* or *three Gods*, is a very considerable article in the writings of several of the ancient christian fathers. Why then do we find nothing of this kind in the age of the apostles? The only answer is, that there then was no occasion for it, the doctrine of the divinity of Christ not having been started." P. 62. He traces a striking resemblance between the character of the Ebionites, as given by the early christian fathers, and that of the Jewish christians at the time of Paul's last journey to Jerusalem. Acts xxi. 20, &c. "So great a resemblance in some things, viz. their attachment to the law, and their prejudices against Paul, cannot but lead us to imagine that they were the same in other respects also, both being equally zealous observers of the law, and equally strangers to the doctrine of the divinity of Christ. And in that age all the Jews were equally zealous for the great doctrine of the *unity of God*, and their *peculiar customs*. Can it be supposed then that they would so obstinately retain the one, and so readily abandon the other? These considerations (and much more might be added to enforce them) certainly affect the credibility

bility of Christ having any nature superior to that of man; and when they are sufficiently attended to (as I suspect they never have been) must shake the Arian hypothesis; but they must be particularly embarrassing to those who, like you, maintain the perfect equality of the Son to the Father."

The sixth letter treats of the personification of the Logos. Under this article, our Author rectifies a mistake of Dr. Horsley's, who had misconceived his meaning. "Those platonizing christians, who personified the Logos, were not Arians; for their Logos was an attribute of the Father, and not any thing that was created of nothing, as the Arians held Christ to have been. It is well known, as Beausobre observes, that they were not Arians, but the orthodox, that platonized."

He shews that the passage in Athenagoras, which Dr. Horsley translated differently from him, does not affect his conclusion from it. "For he evidently asserts, that the Logos was eternal in God only, because God was always λογικος, *rational*, which entirely excludes proper personification. (See Athenagoras, p. 82.) Can reason, as it exists in man, be called a person, merely because man is a *rational* being?"

being?" He says that this is the only one of his authorities that Dr. Horsley has thought proper to examine, and that there are others which he has overlooked so plain and determinate, that it is impossible for him to interpret them otherwise than he has done; as they evidently imply that it depended upon the Father's will that the Logos should have a proper personification, and become a son, with respect to him." He calls upon him particularly to consider the passages he has quoted from Tertullian, which shews how ready the platonizing christians were to revert to the idea of an *attribute* of God in their use of the word Logos. He combats Dr. Horsley's assertion concerning the personal existence of the Logos from all eternity, as contrary to the plainest passages of the Fathers respecting the period of his generation. He charges Dr. Horsley with a total ignorance of what he had asserted, and says p. 72, "The Logos of the Platonists had, in their opinion, always had a personal existence, because Plato supposed creation to have been eternal; but this was not the opinion of the platonizing christians, who held that the world was not eternal; and therefore, retaining as much of platonism as was consistent with

that

that doctrine, they held that there was a time when the Father was *alone*, and without a son; his Logos or reason being all that time the same thing in him that reason now is in man, and of this I have produced abundant evidence.

He produces a curious passage from Justin Martyr, by which it appears that it was the opinion of some in his time, "that the emission of the Logos, as a person, was an occasional thing, and intended to answer particular purposes only; after which it was absorbed into the divine essence again." This opinion our Author thinks probably preceded that of Justin Martyr, and paved the way for it. Concerning it, after quoting the passage, he has the following reflections: p. 75. "We see in this passage in how plausible a manner, and how little likely to alarm men of plain understandings, was the doctrine of the divinity of Christ, as it was first proposed. At first it was nothing more than the *divine power*, occasionally personified (a small step indeed, if any, from pure Unitarianism) and afterwards acquiring permanent personality; but still dependent upon the will of God, from whence it proceeded, and entirely subservient to him; which was very different from what is now conceived

conceived concerning the second person in the Trinity."

The seventh letter contains considerations relating to the doctrine of the Trinity. Our Author here introduces remarks on Dr. Horsley's observations in defence of the Athanasian doctrine, implying a perfect equality in all the three persons. 1. He shews that Dr. Horsley's definition of the doctrine implies a direct contradiction. 2. That his explication of the derivation of the second person from the Father's contemplation of his own divine perfections, is absurd and impossible, but if it could be admitted, a multiplication of divinity without end would be the natural consequence. 3. He proves that the Father is *alone* God from his being the object of prayer, and from our Lord's always addressing him as such. 4. He shews the inutility of the doctrine of the Trinity, one divine person being fully adequate to every purpose that we can conceive. 5. He retorts Dr. Horsley's irony upon himself, and shews that the Socinian interpretations of scripture are the most natural, and agree best with the plainest affirmations of the sacred writers. 6. He says, "that there is nothing that can be called an account of the divine, or even super-angelic super-angelic nature of Christ in the gospels of Matthew, Mark or Luke; and allowing that there may be some colour for it in the introduction of the gospel of John, it is remarkable that there are many passages in his gospel which are decidedly in favour of his simple humanity." He urges strongly this powerful argument, that if the doctrine of the Trinity had been true, it would have been as explicitly declared as that of the Unity is. 7. He affirms, that the apostles could not have continued to call Christ a man simply, after they had been convinced that he was God, and yet they continue to do so in their writings to the last, even in *reasoning* and *argumentation*, without any caveat to prevent their meaning from being misunderstood. 8. If Christ had been God, or the maker of the world under God, he could never have said that *of himself he could do nothing*, that *the words which he spake were not his own*, and that *the Father within him did the works*, &c. 9. He makes light of Dr. Horsley's argument in favour of the Trinity, from some resemblance to it being found in the idolatry of the Heathens and Pagan philosophy, and his considering this in connection with what he imagines he finds in scripture on the subject,

as

as the consent of the latest and earliest revelations. Our Author here puts the following three pertinent questions to his antagonist. "First, if there be so many traces of the doctrine of the Trinity in the heathen philosophy, and in the heathen worship, why are there no more of them to be found in the Jewish scriptures, and in the Jewish worship? Secondly, if there be such traces of the doctrine of the Trinity in the Jewish writings and worship, how came the Jews, in our Saviour's time, and also the body of the Jewish nation to this day, not to discover these traces? Thirdly, if the Jews had once been in the possession of this knowledge, but had lost it in the time of our Saviour, why did not he, who rectified other abuses, rectify this, the most important of them all?"

The eighth letter treats of miscellaneous articles. Our Author objects to Dr. Horsley's improved assertion, that the Ebionites held an unintelligible notion of the exaltation of the nature of Christ after his ascension, and worshipped him as if his nature had been originally divine, and that Theodotus so far surpassed them in his idea of the mere humanity of Christ, as to justify Eusebius in calling him the inventor of the doctrine, &c. He vindicates his translation

lation of a passage in Origen, in regard to the piety of the ancient Unitarians. He acknowledges two mistakes in translating passages from Theophilus, but maintains "that neither Theophilus, nor any person of his age, made a proper *trinity of persons in the Godhead*; for they had no idea of the perfect equality of the second and third persons to the first." He asserts, that the Fathers before the first council of Nice, held, in the most explicit manner, the superiority of the Father to the Son, and refers to the third section of his history for an unanswerable proof of it.

In the concluding letter, our Author refers to some illiberal reflections of Dr. Horsley on his manner of reasoning, his situation as a Dissenter, and Dr. Horsley's charging him with borrowing most of his arguments from Zuicker, whose writings, as they are exceedingly scarce, he had never seen.

The Postcript contains some extracts from Origen referred to in the letters, with notes and observations relating to the subject of them, with some larger articles, viz. The excommunication of Theodotus by Victor, Justin Martyr's account of the knowledge of some christians of low rank, a full and distinct discussion

cussion of the passage in Justin Martyr concerning the Unitarians of his time, of the first author of the doctrine of the permanent personality of the Logos, maxims of historical criticism, with a summary view of the evidence for the primitive christians having held the doctrine of the simple humanity of Christ, most ingeniously drawn up, mutually referring to one another, bringing all the material arguments under a clear and concise view, and exhibiting a criterion by which they ought to be tried: remarks on the article of the Monthly Review for September 1783, in answer to the Author's reply to some former animadversions in that work, before taken notice of.

About the same time (1783) our Author published, *A General View of the Arguments for the Unity of God, and against the Divinity and Pre-existence of Christ, from Reason, from the Scriptures, and from History.* Birmingham, 1783, Price *twopence.* In this valuable little Tract, the Author recites the distinct modifications of the doctrine of the Trinity, and shews that upon any of them there is either no proper *unity* in the divine nature, or no proper *trinity.* He shews from various considerations the

the extreme improbability of the Arian hypothesis. He alleges the most cogent scripture passages in favour of the unity of God and humanity of Christ, arranged under nine distinct heads, with suitable reflections arising from the consideration of them. The summary view of the evidence for the primitive christians having held the doctrine of the simple humanity of Christ, with the maxims of historical criticism by which the particular articles of the said summary may be tried, are here reprinted. This piece, therefore, may be considered as a miniature or compend of Dr. Priestley's ideas with respect to the subjects discussed in it, and from its cheapness and conciseness a very estimable present to those who have not leisure or ability to consult his large publications. It has been re-printed several times, and particularly by the Unitarian society in 1791, who republished it together with the *Appeal* and *Familiar Illustrations*, in one small volume, 12mo.

In the Monthly Review, an uncandid account was given of the *Letters* to Dr. Horsley, and Dr. Priestley was charged by the Reviewer with controversial disingenuity, and mutilating a passage of Justin Martyr quoted by him. This led our Author *once more* to vindicate himself from the aspersions of the Reviewer, whose name had now been discovered, in a small Tract, entitled, *Remarks on the Monthly Review of the Letters to Dr. Horsley; in which the Rev. Mr. Badcock, the Writer of that Review, is called upon to defend what he has advanced in it.* Birmingham, 1784. Our Author shews in his reply to Mr. Badcock, that the words omitted had no relation to the subject for which the passage was quoted, and that they were omitted merely to save himself the trouble of writing so much Greek unnecessarily *. He also points out a gross mistake in Mr. Badcock's translation of the passage. The rest of the Pamphlet treats of the creed of Tertullian, and some miscellaneous articles relating to the controversy.

Notwithstanding our Author's attention was so closely engaged in defending his History of the Corruptions of Christianity, and in making preparations for a large and new work respecting the state of early opinions concerning Christ, he found leisure at this

* It appeared afterwards, and was taken notice of by Dr. Priestley himself, that the passage was really quoted in Greek, and omitted only in the English translation.

time

time for the publication of an excellent devotional composition, entitled, *Forms of Prayer, and other Offices, for the use of Unitarian Societies.* Birmingham, 1783. Besides proper forms for the morning and evening service of the Lord's day, he has here given offices for infant and adult baptism, a form for the celebration of the Lord's supper, addresses to the communicants for a second and third service, a funeral service, prayers for a fast day, an introductory prayer on a day of public thanksgiving, a prayer respecting the present state of Christianity to be used in the morning of Easter Sunday. The Author's object in this publication was to enable Unitarian christians to conduct all the parts of public worship themselves, when deprived of the advantage of a learned ministry, and in the preface and introduction, the most cogent arguments are offered for the necessity of forming such Unitarian societies, and directions given for managing all the different services with propriety and to general edification.

At a pretty advanced period of the year 1784, Dr. Horsley published an answer to our Author's letters addressed to him, entitled, *Letters from the Archdeacon of St. Albans, in reply to Dr. Priestley;*

with

with an Appendix, containing short Strictures on Dr. Priestley's Letters, by an unknown hand. In these letters he declines a regular controversy with Dr. Priestley respecting the doctrine of the Trinity, cavils at some parts of our Author's history which he had passed unnoticed before, and recapitulates the objections contained in his charge. He denies that the clear sense of scripture is in favour of the Unitarians, and insists that Dr. Priestley argues in a circle. He refers to a letter signed *Perhaps* in his Appendix, for an explanation of the word ουτος, in which, after much shuffling, he is obliged to grant that it may be rendered differently from what he has done, though he still thinks his own the most proper translation. He attempts to confute our Author's sense of the phrase *to come in the flesh*, defends his own interpretation of Clemens Romanus, and asserts that the shorter epistles of Ignatius are genuine. He maintains the *difference* between the Ebionites and Nazarenes, criticises some passages of Epiphanius, translates them differently from Dr. Priestley, and asserts that the Nazarenes were no sect of the apostolic age, and that Ebion was not contemporary with St. John. He differs from Dr. Priestley in the interpretation of

two

two passages of Origen, but being aware that his own explication might not stand good, he at last taxes the veracity of Origen, and quotes a passage from Mosheim as follows: "I would not believe this witness upon his oath, vending, as he manifestly does, such flimsy lies." He attempts to controvert Dr. Priestley's maxim, that "whole bodies of men do not soon change their opinion," by appealing to the Dissenters, the whole body of whom formerly, he says, "took their standard of orthodoxy from the opinions of Calvin;" but he adds, "where shall we now find a Dissenter, except perhaps among the dregs of Methodism, who would not think it an affront to be taken for a Calvinist?" He appeals to the epistle of Barnabas as a positive proof that our Lord's divinity was the belief of the very first Christians. Resting the proof of the orthodoxy of the first age upon the epistle of Barnabas, he affirms that Dr. Priestley's two arguments from Hegesippus and Justin Martyr, are overturned. He attempts to combat the testimony of Tertullian in favour of the prevalence of Unitarianism among the lower and unlearned classes of people in his time: and though he cannot help admitting that there is some little foundation for such

an

an inference, yet he attempts by a forced and unnatural construction, and an unfair paraphrase of the words of Tertullian, to abate and enervate their obvious and genuine meaning. He pretends that Dr. Priestley's arguments from Tertullian, Justin Martyr, and Irenæus, to prove that the primitive Unitarians were not heretics, have been confuted by the Monthly Reviewers, and attempts to shew the deficiency of Dr. Priestley's reply, and also to confute his arguments from Clemens Alexandrinus and Jerome. He considers the passage from Athanasius, and endeavours to prove that he speaks of unconverted Jews. He asserts that the divinity of Jesus was acknowledged by the apostles from the time when they acknowledged him for the Messiah. He refers to two places of the gospel as a proof of this assertion, John i. 49, when Nathaniel exclaimed, *Rabbi*, thou art the son of God! thou art the king of Israel, and Luke v. 8, when, after the miraculous draught of fishes it is said of Peter, *he fell down at the knee of Jesus, saying, depart from me, for I am a sinful man, O Lord.* He affirms, that the divinity of Christ was preached from the very beginning by the apostles, that Stephen died a martyr to this doctrine, that

his

his dying ejaculations justify the worship of Christ, that the story of Paul's conversion is another instance of an early preaching of our Lord's divinity, in which Jesus is deified in the highest terms, and that notions of a Trinity, and of the deity of the Messiah, were current among the Jews in the days of our Saviour. He charges Dr. Priestley with a misrepresentation of the Platonic language, denies that the conversion of an attribute into a person was ever taught by the Fathers, and rejects a passage quoted by our Author from Tertullian, and another from Lactantius, as sufficient proofs of the assertion. He attempts a defence of his two suppositions, that the first Ebionites worshipped Christ, and that Theodotus was the first person who taught the Unitarian doctrine at Rome.

Adhering to his declaration not to enter into a regular controversy on the subject of the Trinity, Dr. Horsley gives only a general reply to some parts of Dr. Priestley's seventh letter. Far from entering into the real merits of the question, by meeting his opponent on equal ground, and shewing that the notion of a Trinity in unity implies no contradiction, he takes it for granted that it falls short of a contradiction,

diction, and only contains some difficulties in it that transcend the reach of human understanding. With respect to the article of worship, and the example of our Saviour, he weakly says, that "our Saviour, as a man, owed worship to the Father," and produces the example of Stephen as a sufficient authority to authorize the worship of Christ. He gives insufficient answers to plain passages of scripture alleged by Dr. Priestley, and passes by others altogether. He covers himself with impenetrable mystery, and refers to the Parmenides of Plato for a solution of difficulties. He maintains, however, that what he calls the Catholic faith is supported by the general tenor of the sacred writings, but brings no proof of the truth of this assertion from the scriptures, though he quotes Bishops Bull and Pearson, and Dr. Waterland, for a proper definition of the doctrine of the Trinity, in opposition to an assertion made by Dr. Priestley. He thinks the Unitarian doctrine not well calculated for the conversion of Jews, Mahometans, or Infidels, and, concealing the real state of the case, has put together some unfounded or precarious reasonings of his own to give a colour to the assertion. In conclusion, he gives an account of the progress of his mind in forming his religious principles that does not appear very probable, and in the true spirit of a high churchman, intimates the necessity of a priesthood derived by regular succession from the apostles, and passes a censure upon all voluntary associations of christians who dissent from it. The Short Strictures by an unknown hand, in the Appendix, contain some petty cavils of little moment.

In about three months after the publication of Dr. Horsley's letters, a reply on the part of Dr. Priestley appeared, entitled, *Letters to Dr. Horsley, Part II containing farther Evidence that the Primitive Christian Church was Unitarian.* Birmingham, 1784. In the Preface, which treats of various matters, our Author takes notice of the *Clementines*, "which though properly a *theological romance*, is a fine composition of its kind." Our Author thinks it was written about the time of Justin Martyr, and among other observations concerning it, has the following remark: "Now this writer, whose knowledge of the state of opinions in his time cannot be questioned, would hardly have represented Peter and Clement as Unitarians, if he had not thought them to be such. Nay, it may be inferred from the view that he has given of their principles, that supposing the doctrine of the Trinity to have existed in his time, yet that Peter, Clement, and consequently the great body of christians in the apostolic age, were generally thought to have been Unitarians, as he must have imagined that this circumstance would contribute to the credibility of his narrative."

Our Author, in the beginning of his work, states Dr. Horsley's opinion, and his own contrasted with it, under seventeen different heads, in order to enable his readers to form a clear and comprehensive idea of the *nature* and *extent* of the controversy.

The reply consists of nineteen letters. After an introductory one, our Author, in the second letter, treats of Dr. Horsley's *positive proof*, from the epistle of Barnabas, " that the divinity of our Lord was the belief of the very first christians." He observes, " I am surprised, sir, at the extreme confidence with which you treat this very precarious and uncertain ground; when, to say nothing of the doubts entertained by many learned men concerning the genuineness of this epistle, the most that is possible to be admitted is, that it is genuine *in the main*. For, whether you may have observed it or not, it is most evidently

dently *interpolated*, and the interpolations respect the very subject of which we treat. Two passages in the Greek, which assert the pre-existence of Christ, are omitted in the ancient Latin version of it. And can it be supposed that this version was made in an age in which such an omission was likely to be made?" After quoting the passages in proof of what he has asserted, our Author adds, "The passage on which you lay the chief stress is only in the Latin version, that part of the Greek copy to which it corresponds being now lost; and all the other expressions that you note, are such as an Unitarian will find no difficulty in accommodating to his principles. Can it be thought at all improbable, that if one person interpolated the Greek, another should make as free with the Latin version. Our Author considers the passage from Clemens Romanus at considerable length, and shews that it has no relation to a state of pre-existence, and that so far from proving that Christ was God, it implies the contrary. He thinks the epistles of Ignatius interpolated in the very place that Dr. Horsley refers to, and that the true sense of Dr. Lardner's words, quoted by Dr. Horsley, refers to such an interpolation.

In

In the third letter, he produces two additional passages from Epiphanius, to prove that the Nazarenes held the proper humanity of Christ as well as the Ebionites, and that both these sects, in the opinion of that writer, existed at the time John wrote his gospel. He also produces a passage from Jerom, in which he asserts, that "the doctrine of the Ebionites was then rising, who said that Christ had no being before he was born of Mary." Our Author adds, "This is only one out of many authorities that I could produce for this purpose, and it is not possible to produce any to the contrary." Dr. Horsley had said (p. 27) "As a certain proof that the Ebionites and Nazarenes were two distinct sects, Mosheim observes, that each had its own gospel." In reply, our Author alleges the authority of Mr. Jeremiah Jones, backed by that of Mosheim's translator, to prove, that the gospel of each was the same, and what is of more consequence the opinion of Jerom, who says, "in the gospel used by the Nazarenes and Ebionites, which is commonly called the authentic gospel of Matthew, which I lately translated from Hebrew into Greek," &c. He proves in opposition to Dr. Horsley, p. 22, 23, that the Ebionites did not deny

deny the authority of the prophetical and other books of the Old Testament, and consequently that it is no proof that Hegesippus was not an Ebionite, because he cites the proverbs of Solomon. He says very properly, p. 23, " It is an argument in favour of the identity of the Nazarenes and Ebionites, that the former are not mentioned *by name* by any writer who likewise speaks of the Ebionites before Epiphanius, though the people so called afterwards were certainly known before his time. The term *Ebionites* occurs in Irenæus, Tertullian, Origen, and Eusebius; but none of them make any mention of *Nazarenes*; and yet it cannot be denied, that they must have been even more considerable in the time of these writers, than they were afterwards; for, together with the Ebionites (if there was any difference between them) they dwindled away, till, in the time of Austin, they were *admodum pauci*, very few. Origen must have meant to include those who were called Nazarenes under the appellation of Ebionites, because he speaks of the Ebionites as being the whole body of Jewish christians; and the Nazarenes were christian Jews as well as they. Jerom seems to use the two terms promiscuously; and in the passage of his letter to Austin,

tin, so often quoted in this controversy, I cannot help thinking he makes them to be the same."

Our Author affirms that Dr. Horsley cannot produce any evidence that Theodotus was considered in a worse light by the ancients than by the Ebionites, and thinks his notion of the Ebionites having held an unintelligible exaltation of the mere human nature of Christ after his resurrection, the most improbable of all suppositions. He quotes Epiphanius to prove, in opposition to Dr. Horsley, that the Nazarenes took their rise as a sect after the destruction of Jerusalem by Titus, and considers the passage from Jerom as sufficiently clear to prove, that they were the same people as the Ebionites, and apprehends that it cannot be inferred from Austin's answer to Jerom, that there was any material difference between them. He examines the writings of Grotius as the most respectable of the modern authorities alleged by Dr. Horsley, and quotes a passage from him which contains nothing favourable to Dr. Horsley's sentiments, but afterwards in his Appendix, p. 217, he explains himself farther on the subject, and gives his opinion respecting the sentiments of Grotius, and the little stress that ought to be laid on a passage quoted by that

that Author from Sulpitius Severus respecting the Nazarenes. He points out Dr. Horsley's egregious mistakes in asserting that the generality of the Dissenters had departed from their attachment to Calvinism, and observes that "as they were universally Calvinists at the time of the Reformation, they are very generally so still. The ministers, as might be expected, are the most enlightened, and have introduced some reformation among the common people; but a majority of the ministers are, I believe, still Calvinists."

The fourth letter treats of the supposed orthodox church of Jerusalem, and of the veracity of Origen. Our Author finds no evidence for the existence of such a church of Jewish christians, and considers what Mosheim and Dr. Horsley have advanced on this subject as a mere fiction uncountenanced by any ancient authority: the passage referred to by Mosheim in his ecclesiastical history from Sulpitius Severus not authorising the conclusion. Now though the testimony of that writer were to the purpose, can his authority be compared to that of Origen, when he lived two hundred years later, and at a remote distance from Palestine. Our Author quotes Tille-

mont

mont and Fleury, whose views of this historical fact coincide with his own; defends the veracity of Origen, and intimates that unless Dr. Horsley can make a better apology for himself, than he is able to suggest, he will be considered by impartial persons as a *falsifier of history*, and a *defamer of the character of the dead.*

In the fifth letter, which relates to heresy in the earliest times, our Author re-considers and defends his former interpretation of the phrase *coming in the flesh*, used by the apostle John with respect to Christ. He observes as follows, p. 48. "You say, p. 27, "The attempt to assign a reason why the Redeemer "should be a man, implies both that he might have "been, without partaking of the human nature, and "by consequence, that in his own proper nature he "was originally something different from man; and "that there might have been an expectation that he "would make his appearance in some form above "the human." But it is certainly quite sufficient to account for the apostle's using that phrase *coming in the flesh*, that in his time there actually existed an opinion that Christ was not truly a man, but was a being of a higher order, which was precisely the

doctrine

doctrine of the Gnostics. That before the appearance of the Messiah, any persons expected that he would, or might come, in a person above the human, I absolutely deny."

"A reason," you say, p. 27, "why a man should be a man, one would not expect in a sober man's discourse." But certainly, it was very proper to give a reason why one who was not thought to be properly a man, was really so; which is what the apostle has done. He quotes a passage from Polycarp to prove, that the phrase *coming in the flesh*, is descriptive of the Gnostic heresy only, and not of the Unitarian doctrine also, and recites another from Ignatius, in which he appears to have had the Gnostics in his eye as the only heretics. He finds no reference to the Ebionites in the epistles of Ignatius, except perhaps in the passages which he supposes to have been altered, and produces three other places which are unfavourable to the doctrine of the divinity of Christ.

In the sixth letter, he reviews the sentiments of Justin Martyr, Irenæus, and Clemens Alexandrius, concerning heresy, and considers their censures of it as applicable to the Gnostics, and not to the Unitarians.

The seventh letter gives an account of the state of heresy in the time of Tertullian. In this our Author re-considers at large, the famous passage from that writer relating to the *Idiotæ*, or common unlearned people: the *major pars credentium*, or majority of believers, who held fast to the rule of faith concerning one God, and shuddered at the *œconomy*, or doctrine of the Trinity, when proposed to them. He confutes, in a clear and masterly manner, the sophistry and false comments of Dr. Horsley on the passage, and proves that it plainly asserts, that a very great majority of the unlearned body of Christians in Tertullian's time were Unitarians.

In the eighth letter, Origen's idea of heresy is examined, and several passages from him are produced, to prove that the doctrine of the *Logos*, in the orthodox sense, was not received or understood by the multitude of Christians, who "knew nothing but Jesus Christ and him crucified." Our Author concludes this letter as follows, p. 78, "From all these passages, and others quoted before, especially the *major pars credentium* of Tertullian, I cannot help inferring;

inferring, that the doctrine of Christ being any thing more than a man, who was crucified and rose from the dead (the whole doctrine of the incarnation of the eternal Logos, that was in God, and that was God) was considered as a mere abstruse and refined doctrine, with which there was no occasion to trouble the common people; and it is evident that this class of christians was much staggered by it, and offended when they did hear of it. This could never have been the case if it had been supposed to be the doctrine of the apostles, and to have been delivered by them as the most essential article of christian faith, in which light it is now represented. Such terms as *scandalizare*, *expavescere*, &c. used by Tertullian, and *ταρασσειν*, by Origen, can only apply to the case of some novel and alarming doctrine, something that men had not been accustomed to. In the language of Origen, it had been the *corporeal gospel* only, and not this *spiritual* and *mysterious* one that they had been taught."

In the ninth letter, various passages are produced from Athanasius, Cyril of Jerusalem, Basil and Facundus, to shew that Unitarianism prevailed, particularly

cularly among the common people, in a greater or less degree till the fifth and sixth centuries.

In the tenth letter, our Author shews that it was not merely the opinion of Athanasius, that the apostles used caution or prudential reserve in communicating the doctrine of the divinity of Christ, but that Chrysostom and other Fathers, in several passages which he quotes, represent them as acting in a similar manner; and he justly infers from these acknowledgments, that even in the opinion of these Fathers, at the time of the publication of the gospels, the Christian church was principally Unitarian, believing only the simple humanity of Christ, and knowing nothing of his divinity or pre-existence. From the state of the case as here represented, our Author reasons as follows, p. 101. "From the acknowledgment which these orthodox Fathers could not help virtually making (for certainly they would not do it unnecessarily any more than yourself) that there were great numbers of proper Unitarians in the age of the apostles; it seems not unreasonable to conclude, that there were great numbers of them in the age immediately following, and in their own, and their knowledge of this might be an additional reason for the opinion

opinion that they appear to have formed of that pre-valence in the apostolic age. Would those Fathers have granted to their enemies spontaneously, and contrary to truth, that the Jews were strongly prepossessed against the doctrine of the divinity of Christ, and that the Unitarians were a formidable body of Christians while the apostles were living, if it had been in their power to have denied the facts? The consequence of making these acknowledgments is but too obvious, and must have appeared so to them, as well as it now does to you, which makes you so unwilling to make it after them."

In the eleventh letter, in opposition to Dr. Horsley's assertion, that "the Jews in Christ's days had notions of a Trinity in the divine nature," our Author affirms, that it is clearly supposed by Justin Martyr, and all the Christian Fathers, that the Jews expected only a man for their Messiah. He appeals to the gospels as containing a full confutation of Dr. Horsley's opinion." P. 105. "Inform me then, if you can, how our Saviour could possibly, on your idea, have puzzled the Jewish doctors, as he did, reducing them to absolute silence, by asking them how David could call the Messiah his Lord, when he was his

son

son or descendant. For if they had themselves been fully persuaded, as you suppose, that the Messiah, though carnally descended from David, was in fact the maker and the God of David, and of them all, a very satisfactory answer was pretty obvious." He produces the opinion of the learned Basnage, p. 121, as decidedly against Dr. Horsley on this subject. He considers the passages (one excepted) quoted by Dr. Horsley from the gospels and acts of the apostles in proof of the divinity of Christ, and shews that they are nothing to the purpose. The passage he has omitted is the appearance of our Lord to Saul in his way to Damascus, attended by a light exceeding the brightness of the sun at mid-day, thrice mentioned in the book of Acts. This history, however, carries a sufficient refutation in it to all that Dr. Horsley would infer from it; for our Lord replies to Saul in answer to his question, *I am Jesus of Nazareth, whom thou persecutest.* Acts xxii. 8. A proof of true and proper humanity, but by no means of divinity.

In the twelfth letter, our Author defends his assertion, that the platonizing Fathers held the notion of the conversion of the Logos from an attribute into

a person,

a person, and asserts that Dr. Horsley's pretence that they only meant a *display of powers*, or *projection of energies*, is without foundation in their writings.

The thirteenth letter contains considerations relating to the doctrine of the Trinity, in which for the little that Dr. Horsley has explained himself on the subject, our Author has confuted him well, and demonstrated the absurdity and inconsistency of his notions. As a proof that the scriptures contain the clearest declarations of the divine Unity, he refers Dr. Horsley to 1 Tim. ii. 5. 1 Cor. viii. 6. John xvii. 3. quoted by him at full length in his former letters, but remaining unnoticed by Dr. Horsley.

The fourteenth letter treats of Prayer to Christ. Here our Author shews, contrary to Dr. Horsley's assertion, from various examples in scripture, and that of Polycarp at his martyrdom, that the Father is the great object of prayer in the time of persecution, as well as at other seasons.

In the fifteenth letter, a refutation is given of what Dr. Horsley has advanced with respect to the influence of Unitarian principles in preventing the conversion of Mahometans and Infidels.

In

In the sixteenth letter, Bishop Bull is shewn to have been a defender of damnatory clauses in creeds, and a man of a harsh uncharitable spirit towards Arians and Unitarians.

In the seventeenth letter, the representation that Dr. Horsley has given of the state of Dissenters is considered, with reflections on the penal laws to which the Unitarians are subject.

In the eighteenth letter, our Author vindicates himself from the charge of wilful misrepresentation, and other uncandid insinuations brought against him by his opponents in controversy.

The nineteenth and last letter, treats of various miscellaneous articles, in the first of which our Author acknowledges a small inaccuracy in stating the opinion of Valesius, with respect to the loss of the writings of Hegesippus. What our Author affirmed may however be probably inferred, and it is not unlikely that Valerius might have had it in view, though he has not expressly asserted it. An Appendix follows containing some amendments and additions to the letters.

Dr. Horsley had intimated, in the first of his letters to Dr. Priestley, his intention of appearing no more

more in the controversy. But after an interval of eighteen months, he feels himself disposed to resume his pen, and enter again the field of disputation. This pamphlet appeared under the following title, *Remarks upon Dr. Priestley's second Letter to the Archdeacon of St. Albans, with Proofs of certain Facts asserted by the Archdeacon.* London, 1786.

After several sarcastic remarks upon some parts of Dr. Priestley's second letters, accompanied with many airs of self sufficiency and much unmeaning declamation, he proceeds to the relief of the forlorn church of orthodox Jewish Christians at Jerusalem after the time of Adrian, and to repair its foundations which had been too feebly laid in his former attempts to build it. As a necessary step towards the erection of this fabric, the character of Origen must, at all hazards, be run down, and his veracity called in question. He scruples not to say, p. 24, "that in the particular matter in question Origen asserted a known falsehood." To make good this charge against Origen, he quotes a passage from his second book against Celsus, in which, according to his own exposition of it, Origen seems to distinguish three different kinds of Jewish christians, some who had relinquished the old customs of their ancestors, and two others who retained them, though with different views of their value and necessity, contrary to his former assertion in the same book, in which he avers, "that the Hebrew christians in his time had not abandoned their ancient laws and customs; and that they were all called Ebionites." He farther pretends, that in the next sentence, Origen gives us to understand, though more indirectly, that of these three sorts of Hebrews professing Christianity, those only who had laid aside the use of the Mosaic law, were in his time considered as true Christians. He appears willing also to accuse Origen of prevarication and unfair dealing in his criticism upon the word עלמה, in the same book against Celsus. He affirms, that Epiphanius asserts, "that the Hebrew Christians, after Adrian's settlement of the Ælian colony, returned from Pella, whither they had retired from the distresses of the war, to Ælia. He says he holds the testimony of Origen too cheap to avail himself of his triple division of the Hebrew Christians, to prove the existence of the orthodox sect in his time; and appeals to a passage in Jerom's commentary upon Isaiah, where, in his apprehension, Jerom makes a distinction

distinction between *Hebrews believing in Christ*, and the Nazarenes. He quotes a passage from Orosius, in which that writer says, that the Jews were forbidden to enter Jerusalem, but Christians were permitted to enter it, and from a rescript of Adrian preserved by Justin Martyr in his apology, he infers, that that emperor was not unfavourable to Christians. Resting upon these passages, joined to various glosses of his own upon several Fathers, and various conjectures and suppositions, he thinks he has found sufficient evidence for the existence of a church of orthodox Jewish Christians at Ælia, alias Jerusalem, after the expulsion of the Jews by Adrian, and glories not a little on that account. He pretends that there were five classes of Jewish Christians. Jerom's Hebrews believing in Christ, who were orthodox, and had laid aside the use of the Mosaic law. Two kinds of Nazarenes, both orthodox, and retaining the use of that law, the one of which were less bigoted in their attachment to it than the other. Two sorts of Ebionites denying our Lord's divinity, the one admitting and the other rejecting the miraculous conception. Cheap as he pretends to hold the authority of Origen, he endeavours to avail himself of

that

that authority (p. 60, 61) in making out these distinctions. He attempts to prove his former assertion of the decline of Calvinism among the Dissenters from different facts and circumstances that occurred at the meetings of their ministers in the years 1772 and 1773, when they petitioned parliament for a redress of their grievances. He treats of the doctrines of Calvin, and of the Methodists, and concludes with invectives against Dr. Priestley and his writings. Upon the whole, his pamphlet is a very insufficient reply to Dr. Priestley's second set of letters, and several things of importance are passed over without any notice at all.

A reply on the part of Dr. Priestley soon made its appearance, entitled, *Letters to Dr. Horsley, Part III. containing an Answer to his Remarks on Letters, Part II. To which are added, Strictures on Mr. Howe's Ninth Number of Observations on Books ancient and modern.* Birmingham, 1786.

This reply consists of six letters. The first is merely introductory. In the second letter, which respects the veracity of Origen, our Author insists on the general good character of that ancient writer, and the high improbability of his having given a

false

false testimony in the case of the Ebionites. He observes, p. 6, "Had the testimony of Origen to the Unitarianism of the great body of Jewish Christians not been well founded, it was greatly the purpose of many of the early writers (and particularly of Eusebius, who maintained the novelty of the Unitarian doctrine) to have refuted it. But neither Eusebius, nor any other ancient writer, the most zealous for orthodoxy, and the most hostile to Origen on other accounts, has attempted it. Might it not have been expected of Eusebius in particular, that after he had copied Origen's account of the Ebionites, by dividing them into two classes, just as he had done (viz. some of them believing the miraculous conception, and others not) he would have added that, notwithstanding what Origen had said to the contrary, many of them had abandoned the law of Moses, and were believers in the divinity of Christ? But he has not done any such thing. He therefore must have known that he could not do it, and he was not disposed to tell a wilful lie in the case. Indeed, I am willing to think, that few persons are so abandoned as to be capable of doing this." After suggesting some other arguments in favour of the credibility of the

the testimony of Origen respecting the Ebionites, he quotes the passage at full length, in which Dr. Horsley endeavours to confute him from his own writings, and make his evidence appear contradictory, and has the following remarks upon it.

P. 9. "This contains the whole of your curious reasoning, in which you suppose that Origen, in treating of the same subject, and in continuation of the same argument, has given you this pretence for impeaching his veracity as you have done. But surely this writer, who must have known his own meaning, could not have imagined that he had really contradicted himself in two passages, not in different works, written at different times, or in distant parts of the same work (in which he might have forgotten what he had said in one of the passages, when he was writing the other) but in the same work, the same part of the work, and in paragraphs so very near to each other. And I believe no body before yourself, ever imagined that there was any contradiction in them at all.

"In the former he asserts, in general terms, without making any particular exception, that the Jewish Christians adhered to the customs of their ancestors,

and

and in the latter, which almost immediately follows it, he says that his adversary, who had asserted the contrary, would have said what was more *plausible* (not what was *true*) if he had said that some of them had relinquished their ancient customs, while the rest adhered to them; alluding, perhaps, to a few who had abandoned those customs, while the great body of them had not, which is sufficiently consistent with what he had said before. For inconsiderable exceptions are not regarded in general assertions. It would have been very extraordinary indeed, if *no* Jewish christians whatever had abandoned the rites of their former religion, when, in all ages, some Jews, whether they became christians or not, have done so."

After reasoning farther and to good purpose in this way, he refutes p. 11, 12, 13, what Dr. Horsley says Origen gives us to understand, though more indirectly, that of these three sorts of Hebrews professing christianity, they only who had laid aside the use of the Mosaic law, were in his time considered as true christians. He observes, that the most natural construction of the passage is, that Origen says, "It is no wonder that Celsus should

"be

"be so ignorant of what he was treating when he "classed the Gnostics along with Christians, and did "not even know that there were Israelites who pro- "fessed Christianity, and adhered to the laws of "Moses." He shews p. 13, 14, that the other charge of prevarication brought against Origen in regard to the meaning of a Hebrew word before mentioned, is a mere cavil.

The second letter contains general observations relating to the supposed orthodox church of Jewish christians at Jerusalem, after the time of Adrian. Here our Author assigns five good reasons against the existence of such a church, considers the words of Sulpitius Severus as unfavourable to Dr. Horsley's ideas on the subject, and that even those of Orosius will not authorize his conclusions. He appeals to Eusebius, the oldest writer, who mentions the fact, who says, that after the taking of the city by Adrian, the whole nation of the Jews (παν εθνος, which excludes all distinction with respect to religion) were forbidden even to see the desolation of their metropolis at a distance. He calls in question Dr. Horsley's assertion, that Adrian was not unfavourable to Christians, and after some observations

on

on the subject, he adds, p. 20, "There is, therefore, little reason to think that Adrian was so well disposed to Christianity, as to permit the rebellious Jews to remain in Jerusalem on condition of their embracing it."

In the third letter, he considers the testimony of Epiphanius to the existence of a church of orthodox Jewish Christians at Jerusalem after the time of Adrian. He translates the whole passage, which Dr. Horsley had only imperfectly quoted in English, and it appears from it, compared with the Greek original inserted below, that the return of the Jewish Christians from Pella, mentioned in it, is that return which followed the destruction of Jerusalem by Titus, and therefore the passage is nothing to the purpose for which Dr. Horsley has alleged it: for it cannot be proved from it that these returned Jewish Christians remained at Jerusalem after the expulsion of their nation in general from that city by Adrian. Our Author concludes this letter in these words: "On which side then is the *ignorance*, I say nothing of the *fraud*, of which you suspect me in this business? You must, Sir, dig deeper than you have yet done, for the foundation of this favourite church."

 The

The fourth letter respects the evidence from Jerom in favour of the church before mentioned. Our Author gives the passage at full length in Latin and English, shews the inconsistency of Dr. Horsley's reasoning from it, and thinks, that according to the most probable construction of it, the *Hebrews believing in Christ*, and *the Nazarenes* were the same. But he says, p. 30. 2. "Admitting that Jerom alluded to some difference between the *Hebrews believing in Christ* and *the Nazarenes*, it is far from following, that the former were *completely orthodox*, and the latter not. For the phrase *believing in Christ* is applied by Origen and Jerom to the heretical Jewish Christians.........All the difference between these two descriptions of Jewish Christians that Jerom can be supposed to allude to, is such an one as Origen made of two sorts of Ebionites, viz. one who believed the miraculous conception, and the other who disbelieved it; or that of Justin, viz. of those who would hold communion with the Gentile Christians, and those who would not."

3. "Allowing both that the *Hebrews believing in Christ* and the Nazarenes were different people, and that the former were completely orthodox, it will

not

not follow that there was a church of them at Jerusalem, which is the thing that you contend for."

He considers another passage in Jerom from which Dr. Horsley would infer that some Nazarenes held the doctrine of our Lord's divinity, and acknowledged in Christ the Lord of Hosts of the Old Testament; and shews that the inference is not made by Jerom, nor fairly drawn from his words.

Our Author concludes this article in the following words: "Thus I have considered all the evidence, positive or presumptive, that you have produced for the existence of a church of orthodox Jewish christians at Jerusalem after the time of Adrian. I have particularly considered your five quotations from ancient writers, and do not find that so much as one of them is at all to your purpose. Thus again ends this church of orthodox Jewish Christians at Jerusalem, planted by Mosheim, and destroyed by the too copious watering of the Archdeacon of St. Albans."

The fifth letter contains a few observations on Dr. Horsley's sermon on the miraculous conception.

The

The sixth and last letter respects miscellaneous articles. Our Author maintains his former assertion concerning the prevalence of Calvinism among a great majority of the Dissenters; and in a N. B. subjoined to his preface, he mentions that he hears the subject will be considered by a person who is exceedingly well qualified to inform the public concerning it, and to explain the cause of Dr. Horsley's very gross and palpable mistake. He touches briefly some points of the controversy, and replies to Dr. Horsley's invective against his principles and writings. In the Remarks on Mr. Howe's Ninth Number, our Author replies to that writer who had attempted to prove that the body of the Jews expected a God in their Messiah.

Some time before this third set of letters to Dr. Horsley appeared, our Author had published his large important work, entitled, *An History of Early Opinions concerning Jesus Christ, compiled from original writers; proving that the Christian Church was at first Unitarian.* Birmingham, 4 vols. 8vo. 1786.

In this work, Dr. Priestley has accomplished more than any Unitarian writer had performed before

fore him. It was the object of Whiston, Clarke, and Whitby, and others of like sentiments, to establish the Arian or Semi-arian hypothesis. Little information could therefore be expected from them concerning the corruption of the first simple scheme of Christianity, and the state of Unitarianism in early times. Faustus Socinus, Crellius, and the Polish Unitarians were men of eminent abilities, well versed in sacred criticism, and maintained and defended the Unitarian cause with great skill and dexterity by arguments of reason and scripture, against a numerous host of adversaries, when almost the whole world was in opposition to them. The English Unitarian writers about the time of the revolution, and in the reign of King William, made a conspicuous figure, and left writings behind them which will be long highly prized by those who agree with them in opinion.

Few of these writers, however, either in Great-Britain or on the Continent, turned their attention particularly to the state of things in ecclesiastical antiquity. Zuicker, the ingenious author of *Platonism Unveiled*, a work written originally in French; and an anonymous writer who replied to Bishop Bull's

Defensio

Defensio Fidei Nicenæ, are among the chief of those who have done any thing remarkable in this way. These three writers were no strangers to the *Ebionites* and *Nazarenes* among the Jews, or the *Alogi* among the Gentiles, and have urged some arguments in favour of early Unitarianism with peculiar force. They were not, however, master of the whole mass of evidence on the subject, and probably had never undertaken the laborious task of perusing the whole body of Fathers for four or five centuries after Christ, with a view to throw light on the subject.

At the time Dr. Priestley wrote his History of the Corruptions of Christianity, his knowledge of the subject was not perhaps a great deal superior to that of preceding writers. The merit of the *first part* of that work consists more in the perspicuous and judicious arrangement of facts and circumstances before brought to light, than in any new and fresh accession of materials. Had no violent and hostile opposition been made to the History of the Corruptions, &c. it is probable the Author might have never thought of inquiring much farther; at least, he would have wanted a sufficient motive to stimulate him to encounter the drudgery of turning over the pages of so

many

many voluminous ancient writers, whose obscurity of style, and harshness of diction, are by no means inviting. This last observation is confirmed by our Author himself in writing to Dr. Horsley.

"To yourself, Sir, in particular, the world is indebted for whatever there may be of value in my large *History of early opinions concerning Christ.* For without the link that you put into the chain of *causes and effects,* mechanically operating in my mind, the very idea of that work would not, I believe, have occurred to me*."

To enter into a particular and minute detail of a work so large as this, consisting of so many divisions and sub-divisions, and abounding with such a vast variety of Greek and Latin quotations, would far exceed my limits. I shall therefore content myself with giving the general outline of it.

The work is dedicated to the late Mrs. Rayner, a lady of ample fortune, distinguished by her piety and zeal for rational religion. The large Preface treats of different points relating to the work.

Letters to Dr. Horsley, Part III. p. 47.

The

The Introduction contains a view of the principal arguments against the divinity and pre-existence of Christ. 1. From the general tenor of the scriptures. 2. From the difficulty of tracing the time in which they were divulged. 3. From Christ not being the object of prayer. 4. From the doctrine of the Trinity as implying a contradiction. 5. The nature of the Arian hypothesis is considered, and the proof which is necessary to make it credible. 6. Reasons are proposed for not considering Arians as being properly Unitarians. 7. The argument is stated against the pre-existence of Christ from the materiality of man; and the use, or rather inutility of the doctrine of the Trinity is considered.

The first book contains the history of opinions which preceded the doctrine of the divinity of Christ, and which prepared the way for it.

Chap. 1. Of those who are called Apostolical Fathers.

2. Of the Principles of the Oriental Philosophy.

3. Of the Principles of the Christian Gnostics.

The

The particular tenets of the Gnostics are detailed in eleven different sections.

Chap. 4. The Gnostics were the only Heretics in early times.

Sect. 1. Of Heresy in general. 2. Of Heresy before Justin Martyr. 3. Of Heresy according to Justin Martyr. 4. Of Heresy according to Irenæus. 5. Of Heresy according to Clemens Alexandrinus, Tertullian, Origen, and Firmillian. 6. Of Heresy in a later period.

Chap. 5. Of the Apostles Creed as a guard against Gnosticism.

7. A View of the Principles of the later Platonists.

Sect. 1. The Doctrine of the later Platonists concerning God and Nature. 2. Of the Doctrine of the Platonists concerning the Union of the Soul with God, and General Observations.

VOL. II.....BOOK I.

The History of Opinions which preceded the doctrine of the divinity of Christ, and which prepared the way for it, continued.

Chap. 8. Of the Platonism of Philo.

Book II.

Containing the History of the Doctrine of the Trinity.

Chap. 1. Of Christian Platonism.

2. Of the Generation of the Son from the Father.

Sect. 1. The Doctrine of the Platonizing Fathers concerning the Generation of the Son as the second person in the Trinity, stated.

Sect. 2. Authorities for this opinion from Justin Martyr to Origen. 3. Authorities from Origen and other writers subsequent to him; with an account of other attributes of the Fathers, besides that of wisdom, which Christ is said to have been.

Chap. 3. The Defence of the preceding doctrine by the Fathers.

Sect. 1. The Generation of the Son from the Father, illustrated by the uttering of words. 2. The Generation of the Son from the Father, illustrated by the prolation of a branch of a tree from the root, &c. 3. Why only one son was generated, the objection of generation implying passion considered, and why the Son and Holy Spirit did not generate. 4. Whether the generation of the son was in in time, and also whether it was a voluntary or involuntary act of the Father.

Chap. 4. The inferiority of the Son to the Father, shewn to have been the doctrine of all the Antenicene Fathers.

5. Of the power and dignity of Christ as the pre-existing Logos of the Father.

6. Christ, beside being the Logos of the Father, was thought to have a proper human soul.

7. Of the Union between the Logos, and the soul and body of Christ, and their separate properties.

Sect. 1. Of this Union in general. 2. Of the Ignorance of Christ concerning the Day of Judgment. 3. Opinions concerning the body of Christ.

Chap. 8. Of the Use of the Incarnation, and the objections that were made to the doctrine.

9. Of the Controversy relating to the Holy Spirit.

Sect. 1. Opinions concerning the Holy Spirit before the Council of Nice. 2. Opinions concerning the Holy Spirit after the Council of Nice. 3.

Of

Of the proper office of the spirit with respect to the offices of the Father and the Son. 4. Of the arguments for the Divinity of the Holy Spirit.

Chap. 10. Of the Doctrine of the Trinity after the Council of Nice.

Sect. 1. The doctrine of the perfect equality of all the persons in the Trinity. 2. Of the New Language introduced at and after the Council of Nice. 3. Illustrations of the doctrine of the Trinity.

Chap. 11. Of the arguments by which the doctrine of the Trinity was defended.

Sect. 1. Arguments from the Old Testament. 2. Arguments from the New Testament. 3. Answers to Objections.

VOL. III....BOOK III.

Introduction.

Chap. 1. That the Jews in all ages were believers in the Divine Unity.

Sect. 1. The fact acknowledged by the Christian Fathers. 2. Of the reasons why, according to the Christian Fathers, the doctrine of the Trinity was not discovered to the Jews. 3. The sentiments of the Jews, as expressed by themselves, on the subject. 4. Of the Jewish angel Metatron, &c.

Chap.

Sect. 1. The Acknowledgments of the Christian Fathers that John was the first who taught the doctrines above mentioned. 2. Reflections on the subject.

Sect. 1. Presumptive evidence that the majority of the Gentile Christians in the early ages were Unitarians. 2. Direct evidence in favour of the Gentile Christians having been generally Unitarians.

Chap.

Chap.

BOOK IV.

Of some controversies which had a near relation to the Trinitarian or Unitarian doctrine.

Chap. 1. Of the Arian Controversy.

Sect. 1. Of the antecedent causes of the Arian doctrine. 2. Of the tenets of the ancient Arians. 3. The arguments of the ancient Arians. 4. Of the arguments of the Orthodox against the Arians. 5. General observations against the Arian controversy.

Chap. 2. Of the Nestorian controversy.

3. An account of the Priscillianists and Paulicians.

CONCLUSION.

Sect. 1. A connected view of all the principal articles in the preceding history. 2. An account of the remains of the Oriental or Platonic philosophy, in modern systems of Christianity. 3. Maxims of historical criticism. 4. A summary view of the evidence for the primitive christians having held the doctrine of the simple humanity of Christ. 5. Some of the uses that may be derived from the consideration of the subject of this work. 6. Of the present state of things with respect to the Trinitarian and Arian controversies. Articles omitted to be inserted in

 their

their proper places. An Appendix, containing the remarks of the Author's friends on the work, with corrections and emendations.

From the summary view of the contents of this work before given, a sensible reader unacquainted with the nature of it, will be able to form a better idea of its variety and extent, and the connection and coherence of its parts, than by any imperfect abstract I could have made of it in a short compass. Every article in it is supported by quotations from ancient ecclesiastical writers in Greek and Latin, which are inserted in the notes below, and either translated or the substance of them given in the body of the work. These translations, as our Author informs us in his Appendix, amount to about eighteen hundred. To compose a work of this kind, as our Author did, from original authorities: to inspect so many ancient writers, to select from them the necessary passages, and arrange them in that just and proper order in which they now appear, must have been a task of immense labour, and yet our Author performed it in less than the space of three years. The most important places of the first and second parts of our Author's correspondence with Dr. Horsley are here inserted under their

their proper heads, though without mentioning the name of that writer, and very strong and powerful arguments are offered to prove the general prevalence of Unitarian principles in the first ages of the Christian church, and the rise and progress of the Trinitarian and Arian systems, are very naturally and probably accounted for. In my apprehension, the general plan and execution of this work will long do honour to Dr. Priestley's memory, and have a just value set upon it by all who cherish and embrace Christianity in its genuine and original simplicity.

The publication of this last great work, connected with our Author's preceding controversy with Dr. Horsley, brought several new writers into the field. Some of these, however, threatened more than they performed, and none of them entered largely and distinctly into the controversy in all its parts. Our Author published three replies to these writers, of which we shall here give a brief account. The first is entitled *Defences of Unitarianism for the year* 1786, *containing Letters to Dr. Horne, Dean of Canterbury; to the young men who are in a course of education for the Christian ministry, at the Universities of Oxford and Cambridge; to the Rev. Dr. Price; and*

and to the Rev. Mr. Parkhurst, on the subject of the person of Christ. Birmingham, 1788. After considering in the first letter to Dr. Horne, an accusation brought against him of having charged the defenders of the doctrine of the Trinity with ignorance or insincerity, our Author, in the second letter, treats of the argument from antiquity, and of Dr. Horsley's services with respect to it. He proposes thirteen questions to be answered by Dr. Horne, with respect to different points of the controversy in which he apprehends Dr. Horsley has failed in his proof. The remaining three letters treat of the interference of civil power in matters of religion, of some particular arguments for the doctrine of the Trinity, and of miscellaneous articles.

The letters to the young men, &c. contain discussions on the following topics: Subscription to articles of faith. The study of the doctrine of the Trinity. The difficulties attending an open acknowledgment of truth. Animadversions on Dr. Purkis's Sermon. Mr. Jones's Catholic doctrine of the Trinity.

In the twelve letters to Dr. Price, the arguments proposed by that celebrated writer and excellent christian,

christian, in his sermons in favour of the Arian hypothesis, are distinctly and particularly considered, and replied to with great ability.

The letter to Mr. Parkhurst contains observations on a treatise of that writer, entitled, *The Divinity and Pre-existence of Christ demonstrated from Scripture, in answer to the first section of Dr. Priestley's Introduction to his History of Early Opinions concerning Jesus Christ, together with strictures on some other parts of that work.*

The second reply is entitled, *Defences of Unitarianism for the year* 1787, *containing Letters to the Rev. Dr. Geddes, to the Rev. Dr. Price, Part II. And to the Candidates for Orders in the two Universities, Part II. Relating to Mr. Howe's Appendix to his fourth Volume of Observations on Books, a Letter by an Under-Graduate of Oxford, Dr. Croft's Bampton Lectures, and several other publications.* Birmingham, 1788.

Dr. Geddes had published a small pamphlet in 1787, containing a letter to Dr. Priestley, in which he endeavoured to prove by one prescriptive argument, that the divinity of Jesus Christ was a primitive tenet of Christianity. This prescriptive argument,

ment, he says, is "the formal decision of the Nicene council;" and he asks Dr. Priestley "whether "he thinks it in the smallest degree probable, that "three hundred and eighteen of the principal pas-"tors in the Christian church, convoked from the "three parts of the then known world, could possi-"bly combine to establish a doctrine different from "that which they had hitherto taught their respec-"tive flocks, and which they had themselves receiv-"ed from their predecessors in the ministry."

Our Author addressed four letters to Dr. Geddes. In the first, he shews that the unity of God is declared in the clearest and most decisive manner in the scriptures. In the second, he assigns seven good reasons why the decision of the Nicene council cannot be considered as a fair expression and declaration of the general sentiments of the Christian church, and consequently Dr. Geddes's argument grounded on that decision, is fallacious and inconclusive. In the third, he shews the prevalence of Unitarianism among the great body of Christians in all the periods preceding the council of Nice, and even at the time, and after, that council was held: and in the fourth and last letter, he affirms, in opposition to Dr. Ged-

des,

des, that there can be no *kinds* or *degrees* of divinity, and that if Christ be not God in the supreme and superlative sense of that word, he cannot be considered as God at all. He invites Dr. Geddes to a farther discussion of the controversy, but this invitation he did not embrace.

Dr. Price having, in an Appendix to his Sermons, candidly stated some of the most important of Dr. Priestley's arguments against the Arian hypothesis, and in some places made remarks upon them, our Author, in seven letters, pursues the subject with him with the same acuteness and ability as before; and in this, as well as the former part, has suggested some very powerful arguments both from scripture and reason against the Arian notion of Christ's pre-existence, and his having acted in the creation and formation of the world.

There are eight letters addressed to the candidates for orders; in the five first of which our Author replies to Mr. Howe's uncandid insinuations respecting himself, and his misrepresentation of the doctrine of the Ebionites and other ancient sects. He recapitulates several passages from ancient writers before quoted in other publications, to evince what

what the true tenets of the Ebionites were, and points out the true meaning of a passage in Tertullian, and another in Epiphanius, quoted and misapplied by Mr. Howes.

The three remaining letters contain replies to Mr. Madan and other writers, the letter respecting subscription, &c. addressed to the Author by an Under-graduate, and Dr. Croft's Bampton lectures, in which the young candidates are admonished and guarded against the fallacies of these writers, and exhorted to a steady resistance of all unjust and unreasonable impositions in matters of christian faith.

The third and last reply bears the title of *Defences of Unitarianism for the years* 1788 *and* 1789, *containing Letters to Dr. Horsley, Lord Bishop of St. Davids; to the Rev. Mr. Barnard, the Rev. Dr. Knowles, and the Rev. Mr. Hawkins.* Birmingham, 1790.

Dr. Horsley, after having kept silence nearly three years, was prevailed upon at last (as he informs his readers) by the solicitation of his friends, to republish his former controversial tracts at Gloucester, 1780. To these he added a preface, notes, and six supplementary disquisitions. The preface contains a brief a brief and partial view of the state of his controversy with Dr. Priestley, and a declaration on his part that he had not, and did not intend to read Dr. Priestley's History of Early Opinions. The Disquisitions are employed on the following subjects. 1. Of the Phrase "coming in the flesh," as used by Polycarp in his epistle to the Philippians. 2. Of the passage in Tertullian respecting the Unitarians, and his use of the word *Idiota*. 3. Of the sentiments of Irenæus with respect to the Ebionites. 4. Of the sentiments of the Fathers and others concerning the eternal organization of the Son in the necessary energies of the paternal intellect. 5. Of Origen's want of veracity. 6. Of St. Jerome's orthodox Hebrew Christians. These dissertations, though highly laboured, and composed no doubt with much deliberation in the course of three years, are far from being conclusive or convincing with respect to the subjects of which they treat. The only one of them in which he appears to have gained any advantage, is the third; and that only respects the opinion of Irenæus about the Ebionites, whether in that Father's judgment they were heretics or not. Dr. Horsley has been at pains to collect a number of passages from that writer concerning this sect, from which it appears he had an unfavourable idea of them, and in one of which he expressly calls them heretics.

Dr. Priestley's reply consists of ten letters. In the first, he considers and properly exposes Dr. Horsley's attempts to depreciate his antagonist. In the second, he replies to the charge of want of candour. In the third, he renews the subject of borrowing from Zuicker, and relates a circumstance from which it seems fair to conclude, that notwithstanding all that Dr. Horsley had said concerning that writer, he had never seen his book at all. Dr. Priestley having had a copy of Zuicker's work sent him by a foreign correspondent, gives here a brief account of it. The fourth letter treats of the damnatory clause in the Athanasian creed. In the fifth letter, our Author defends his interpretation of the phrase, *coming in the flesh*, in answer to Dr. Horsley's first supplementary disquisition. In the sixth letter, he re-considers briefly the passage from Tertullian, and the meaning of the word Idiota, and exposes the laboured sophistry of his antagonist in his second disquisition. In the seventh letter, he considers the opinion of Irenæus concerning heretics, and

acknowledges

acknowledges that he had overlooked a passage quoted by Dr. Horsley from that writer, in which the Ebionites are called by that name; but he affirms, that according to the account of the principles of early heretics given by Irenæus, that to have been consistent with himself, he ought not to have considered the Ebionites as heretics. He regards, however, the opinion of Irenæus, as of no consequence to the argument, and would have produced the passage himself if it had occurred to his perusal.

The eighth letter respects Dr. Horsley's notion of the origin of the Son from the Father's contemplation of his own perfections, in answer to his fourth Disquisition. Here our Author shews, by express quotations from Tatian, Theophilus, Clemens Alexandrinus, Hippolytus, Tertullian, Novatian, Lactantius, Eusebius, and Athanasius, that this notion of Dr. Horsley's was incompatible with the idea that these Fathers had of the generation of the son from the Father's will and intention, and that all Dr. Horsley's authorities for it are derived from modern theological writers.

In the ninth letter, our Author, by a recapitulation of known facts and circumstances, defends the

veracity

veracity of Origen, and overturns the precarious suppositions of Dr. Horsley in regard to the existence of a church of orthodox Jewish christians at Jerusalem. He concludes this letter as follows. "To shew that I am not ambitious of having the last word, except where I have something of importance to add, I also freely submit to our readers what your Lordship has added in your sixth dissertation concerning *Jerom's orthodox Hebrew Christians*, in answer to the *fourth* of my *third set of Letters*. That the Ebionites and Nazarenes were only two names for the same set of people, and that they were all, as far as we know, believers in the simple humanity of Christ, I have abundantly proved in my *History of early opinions concerning Jesus Christ*; and certainly your Lordship's not chusing to look into that work, cannot be called an *answer* to it. Till I see something at least *plausibly* advanced in answer to what I have there alleged, I shall think it unnecessary to say any thing farther on the subject."

The tenth letter contains reflections on Dr. Horsley's insolent and uncandid method of conducting the controversy, and his making no acknowledgments even with respect to these points of which he has not attempted

attempted to renew the defence. He calls upon him and other champions in the establishment, to come forth again in support of their cause, and in order to stimulate them the more effectually, he quotes and translates a passage from the *Prolegomena* of Dr. Cave's *Historia Literaria*, in which that writer endeavours to animate the clergy of the church of England to defend her doctrines against the Unitarian writers of his time. He observes, that it has been said that Dr. Horsley has already been rewarded with a bishopric for his former services in the cause of orthodoxy, and that new exertions may still raise him higher in the scale of ecclesiastical preferment.

Our Author, in six letters addressed to Mr. Barnard, a Roman Catholic writer, vindicates Unitarians and their cause from his aspersions, refutes his arguments in favour of the Trinity from the scriptures and from ecclesiastical antiquity, and replies to his vindication of Dr. Geddes's account of the council of Nice, and the prescriptive argument founded upon it; and in a letter to Dr. Knowles, he confutes the weak arguments and reasonings of that writer.

There are eight letters addressed to Mr. Hawkins, who had formerly been a Roman Catholic, and had lately come over to the communion of the church of England, though as it too evidently appears from Dr. Priestley's quotations from his publication, that he was far from being completely satisfied with the doctrines of that church. The five first letters treat chiefly of subscription to human articles of faith, in regard to which Mr. Hawkins has recourse to various methods and expedients to satisfy himself, which are very properly animadverted upon by Dr. Priestley. The three last letters respect the doctrine of the Trinity, and points connected with it, in which Mr. Hawkins's attempts to reconcile that doctrine to the scriptures, to reason, and his own conscience, are well confuted and exposed by Dr. Priestley.

An Appendix follows this Tract, containing an account of no less than *fourteen* senses, in which the subscription of the thirty-nine articles of the church of England has been vindicated. These different senses Dr. Priestley says were collected by an *ingenious friend* of his, at that time living in England. This gentleman is now known to have been Thomas Cooper,

Cooper, Esq. at present occupying the respectable station of a Judge in Pennsylvania.

We must now go back a little, and give an account of some works that Dr. Priestley published during the time he was engaged in the controversies before mentioned, the thread of which we were unwilling to interrupt by inserting any thing foreign to the subject of them. In 1784, the Theological Repository was revived, and three additional volumes were published in that and succeeding years. The Essays written by Dr. Priestley himself in these three volumes, under the signatures of *Pamphilus*, *Hermas*, *Pelagius*, *Beryllus*, *Biblicus*, *Josephus*, *Ebionita*, *Photinus*, and *Scrutator*, are too numerous to be enlarged upon; we shall, therefore, only give their titles as follows, viz.

15. An

34. Of

In 1786, our Author published at Birmingham, *Letters to the Jews; inviting them to an amicable Discussion of the Evidences of Christianity*. A second edition, with some additions, appeared in 1787. This sprightly animated piece was well calculated to make an impression on the Jews, if their inveterate prejudices against Christianity would permit them to listen with candour to any thing that can be said in favour of it. It consists of five letters, viz. 1. Of the peculiar Privileges of the Jewish Nation, and the Causes of their Prejudices against Christianity. 2. Of the present dispersed and calamitous state of the Jewish nation. 3. Of the Historical Evidences of the divine mission of Christ. 4. Of the Doctrine concerning the Messiah. 5. Miscellaneous Observations, and Conclusion. David Levi, a Jew, having published an answer to this piece of Dr. Priestley's, our Author addressed a second set of letters to the Jews, seven in number, in which Mr. Levi's objections are particularly considered and obviated.

About

About the same time, our Author published, *Discourses on various subjects, including several on particular occasions.* Birmingham, 1787. The subjects of these discourses are as follows.

1. A serious attention to Christian duties; a sermon preached before the congregation of Protestant Dissenters, at Mill-hill chapel, in Leeds, May 16, 1773, on occasion of resigning the pastoral office among them, before noticed.

2. The Uses of Christian societies; a sermon preached Dec. 31, 1780, at the New Meeting, in Birmingham, on undertaking the pastoral office in that place.

3. The proper Constitution of a Christian Church, considered in a sermon preached at the New Meeting, in Birmingham, Nov. 3, 1782; to which is prefixed, a prefatory discourse, relating to the present state of those who are called rational Dissenters.

4. The Importance and Extent of Free Inquiry in matters of religion; a sermon preached before the congregations of the Old and New Meeting of Protestant Dissenters at Birmingham, Nov. 5, 1785.

5. The Doctrine of Divine Influence on the Human Mind; considered in a sermon preached at the ordination

ordination of the Rev. Thomas and John Jervis, in 1779.

6. Two Discourses. 1. On Habitual Devotion. 2. On the Duty of not living to ourselves; both preached to assemblies of Dissenting Ministers, and published at their request.

7. Of the Danger of Bad Habits.

8. The Duty of not being ashamed of the Gospel.

9. Glorying in the Cross of Christ.

10. Taking the Cross and following Christ.

11. The Evidence of Christianity from the Persecution of Christians.

To the Discourse on the Nature and Extent of Free Inquiry, when first printed by itself, were annexed, Animadversions on some Passages on Mr. White's Sermons at the Bampton Lectures; Mr. Howe's Discourse on the Abuse of the Talent of Disputation in Religion; and a Pamphlet, entitled, "Primitive Candour," with notes, and the History of a Calumny, re-printed from the St. James's Chronicle of Jan. 21, 1773.

Besides the sensible and valuable discourses contained in the volume before mentioned, our Author

published

published occasionally, from 1788 to 1791, several discourses of particular excellence, which have never been collected, viz.

1. A Sermon on the subject of the Slave Trade; delivered to a Society of Protestant Dissenters, at the New Meeting, in Birmingham; and published at their request. Birmingham, 1788.

2. The Conduct to be observed by Dissenters, in order to procure the Repeal of the Corporation and Test Acts. Recommended in a sermon, preached before the congregations of the Old and New Meetings, at Birmingham, Nov. 5, 1789. Printed at the request of the Committee of the Seven Congregations of the three denominations of Protestant Dissenters, in Birmingham.

3. Reflections on Death; a sermon, on occasion of the death of the Rev. Robert Robinson, of Cambridge, delivered at the New Meeting in Birmingham, June 13, 1790. And published at the request of those who heard it, and of Mr. Robinson's family. Birmingham, 1798.

4. A View of Revealed Religion; a sermon, preached at the ordination of the Rev. William Field of Warwick, July 12, 1790. With a Charge, delivered

delivered at the same time, by the Rev. Thomas Belsham. Birmingham, 1790.

5. The proper Objects of Education, in the present State of the World: represented in a discourse, delivered on Wednesday, April 27, 1791, at the Meeting-house in the Old Jewry, London; to the Supporters of the New College at Hackney. To which is subjoined a Prayer, delivered at the same time, by Thomas Belsham. 2d edit. London, 1791.

6. A Discourse on occasion of the death of Dr. Price; delivered at Hackney, on Sunday, May 1, 1791. London, 1791.

To this Discourse is annexed, A short Sketch of the Life of Dr. Price, with an account of all his publications.

7. The Evidence of the Resurrection of Jesus considered, in a Discourse first delivered in the Assembly-room, at Buxton, on Sunday, September 19, 1790. To which is added, An Address to the Jews. Birmingham, 1791.

8. The Duty of Forgiveness of Injuries; a Discourse intended to be delivered soon after the riots in Birmingham. Birmingham, 1791.

9. A particular

9. A particular Attention to the Instruction of the Young recommended, in a discourse delivered at the Gravel-pit Meeting, in Hackney, Dec. 4, 1791, on entering on the office of Pastor to the congregation of Protestant Dissenters, assembling in that place. London, 1791.

In 1787, our Author published, *A Letter to the Right Hon. William Pitt, on the Subjects of Toleration and Church Establishments, occasioned by his Speech against the Repeal of the Test and Corporation Acts, on Wednesday the 28th of March*, 1787. London, 1787.

The impolicy and injustice of the test and corporation acts, the necessity of repealing the penal laws in force against Unitarians, the evils attending the ecclesiastical establishments of England and Ireland, the unscriptural doctrines maintained in them, the impropriety of excluding Dissenters from the Universities, with other topics of a similar nature, are here laid before the minister, and insisted upon with much spirit and propriety; in order to give him clearer and juster ideas on these subjects, than he appeared to be possessed of, when he delivered his speech above mentioned, in the hearing of our Author.

thor. The attempt was laudable on the part of Dr. Priestley, but like other attempts of a like kind, attended with no good effect. The voice of truth is too feeble to affect the ears of an unfeeling statesman, or to make a favourable impression on his callous and obdurate heart.

The following year our Author re-published, *An History of the Sufferings of M. Louis de Marolles, and M. Isaac le Fevre, upon the Revocation of the edict of Nantz. To which is prefixed, a General Account of the Treatment of the Protestants in the Gallies of France. Translated from the French about the beginning of this century.* Birmingham, 1788.

The sufferings of these two Protestant martyrs in the gallies and prisons of France were very severe, of long duration, and supported with the greatest constancy, patience, and meekness. Dr. Priestley appears to have been greatly affected by the perusal of this narrative, and thought it highly worthy of republication. He has prefixed to it a preface full of pious and instructive sentiments, which will be read with pleasure, as well as the work itself, by those who have a proper conception of Christian magnanimity, and patient suffering for the sake of conscience.

The attempts made by the Dissenters to procure the repeal of the test and corporation acts, the meetings that were held for that purpose, the part that Dr. Priestley took in these proceedings, with the general strain and spirit of his writings, awakened the jealousy and excited the resentment of Mr. Madan and Mr. Burn, two clergymen of the church of England, residing in Birmingham at the same time with Dr. Priestley. Mr. Madan attacked Dr. Priestley first from the pulpit, and then from the press, and Mr. Burn endeavoured to refute some of his late writings. To guard the minds of the people of Birmingham against deception, and to prevent them from conceiving unjust and ill-founded prejudices against the Dissenters in general, or himself and his Unitarian friends in particular, our Author thought it necessary to address them in a series of letters published at short intervals, in five parts, from March to June 1790, and afterwards re-published, joined with letters to Mr. Burn, with some additions and corrections. They are entitled, *Familiar Letters, addressed to the Inhabitants of Birmingham, in refutation*

of

of several Charges advanced against the Dissenters and Unitarians, by the Rev. Mr. Madan. Also, Letters to the Rev. Edward Burn, in answer to his, *on the Infallibility of the Apostolic Testimony concerning the Person of Christ. And Considerations on the differences of opinion among Christians, which originally accompanied the reply to the Rev. Mr. Venn.* 2d Edit. Birmingham 1790.

It appears from the Preface to the second edition, that these Familiar Letters, &c. had a more extensive circulation than most of Dr. Priestley's publications. They are twenty-two in number, written with great ability, and occasionally enlivened with strokes of wit and pleasantry. The first part, (including Letters 1, 2, 3), vindicates the public meetings of the Dissenters from having any seditious tendency, and produces proof from history and recent facts, that they have not been such enemies to monarchy as Mr. Madan has represented. The inconclusiveness of Mr. Madan's reasoning is demonstrated from a variety of considerations.

Part 2, (Letters 4, 5) respects the corporation and test acts, the defeat of the Dissenters in the House

of

of Commons, March 2, 1791, and the conduct of the clergy in procuring it.

Part 3, (Letters 6, 7, 8) treats of a Complete Toleration of Religious Establishments in general, and Remarks on what Mr. Madan has advanced on this subject.

Part 4, (Letters 9 to 16) contains our Author's account of a rude letter from Mr. Madan, treats of Mr. Madan's Apology for his treatment of the Dissenters, of his farther arguments to prove that the Dissenters are unquestionably republican, and of the decision of the House of Commons against the Dissenters, of the ecclesiastical constitution of Ireland, of a charge of being fond of controversy, of the principles of the church of England and subscription to its articles, &c. and of Mr. Madan's idea of Unitarianism.

Part 5, (Letters 17 to 22) gives an Account of Unitarian principles supported by scriptural authorities, treats of Mr. Burn's letters in answer to the Author, contains a short history of the Dissenters, and an account of their general principles, treats of the situation of the clergy of the established church, and of calumnies contained in a Pamphlet, entitled,

Theodosius,

Theodosius, with a conclusion. A postscript is added, containing an account of the Author's intercourse with the late Mr. Badcock.

The Letters to Mr. Burn (six in number) treat of the principle of Mr. Burn's objections to the Author's reasoning concerning the person of Christ; of the argument for the divinity of Christ from Heb. i. 8.; of the reason for appealing to Early Opinions concerning the person of Christ; of the Doctrine of Inspiration; of the immoral consequences of the Author's opinions, and conclusion. An account has been given before, of the Considerations on differences of opinion among Christians.

The same year our Author published, *Remarks on two Letters, addressed to the Delegates from the several Congregations of Protestant Dissenters, who met at Devizes, on Sept.* 14, 1789.

These remarks were annexed to a short but sensible Pamphlet, entitled, *The Spirit of the Constitution and that of the Church of England compared*, composed by another hand, and accompanied by a spirited and suitable Preface written by our Author. The Preface and Remarks are without Dr. Priestley's name, but he acknowledges himself the Author

of

of both in a note subjoined to the Preface of the Familiar Letters, &c. re-published with some additions and corrections in 1790. Several quotations are introduced into these *Remarks* from the *Two Letters*, &c. before mentioned. The sophistry employed by the writer in order to justify the continuance of the corporation and test acts, is refuted by our Author, sometimes directly, and other times by shewing the dangerous or absurd consequences that would result from similar maxims and positions being applied to other subjects. Perhaps a fuller and clearer confutation of the unjust and arrogant claims of high churchmen, can no where be found in an equally short compass, and comprehending so many particulars as the following, viz.

Introduction.

Section 1. Of the Dissenters not having a Right to complain of not being appointed to offices, to fill which no Person can pretend to have a right.

Sect. 2. Of the Dissenters incapacitating themselves for civil offices.

Sect. 3. Of Danger to the State from employing Sectaries.

Sect. 4.

Sect. 4. Of the Dissenters being Enemies to the Constitution.

Sect. 5. Of the Exclusion of Dissenters from Civil Offices by the Church, on the Principle of Self-defence.

Sect. 6. Of the Necessity of an Ecclesiastical Establishment.

Sect. 7. Of the State of Ireland with respeet to the Test Act.

Sect. 8. Of the Policy of the Church of England with respect to the Measure proposed.

All these topics are discussed with ability in the space of 26 pages, and sometimes with a proper seasoning of well-applied drollery.

The researches our Author had made into the state of things in the first ages of Christianity, and his frequent perusal of the Fathers and other ecclesiastical writers for that purpose, very naturally led him to think of writing a general history of the Christian church, and qualified him in some measure for the execution of it. Ecclesiastical history is indeed a beaten field, and has been frequently traversed both by Protestant and Roman Catholic writers. Besides the general histories of Bavorius, Spanheim, Du Pin, Tillemont,

Tillemont, Fleury, Mosheim, &c. the authors who have treated of particular periods or countries, are almost innumerable. Notwithstanding there was still room for the labours and exertions of an ingenious writer like Dr. Priestley. However careful the authors before mentioned might be in collecting and arranging facts, their theological prejudices in a manner necessarily led them to pass over some circumstances slightly, or give a partial account of them. What had been omitted, or incompletely executed by other ecclesiastical historians, our Author has endeavoured to supply in his ingenious work, the first part of which is entitled, *A General History of the Christian Church, to the Fall of the Western Empire.* 2 vols. 8vo. Birmingham, 1790.

Our Author, disapproving of the common division into centuries, has divided the whole time from the birth of Christ to the fall of the Western Empire, in A. D. 475, into thirteen periods. Under each of these periods, the most material facts and circumstances respecting the progress of Christianity, and the difficulties and persecutions it had to encounter, are distinctly related; and, at proper intervals, an account is given of the state of the Jews, the

rise

rise of sects and parties with the controversies occasioned by them, the early ecclesiastical writers, and such civil transactions as had a necessary connection with the history of the church.

The transactions that occurred during the public ministry of our Lord, the propagation of Christianity as recorded in the acts of the Apostles, and such matters of fact as could be collected and inferred from the epistles, are very agreeably told in the first place. Afterwards, the rapidly increasing spread and progress of Christianity, is particularly traced through the reigns of the Emperors Trajan, Adrian, &c. down to Decius and Dioclesian. Our Author has carefully noted the period when the primitive purity of evangelical doctrine began first to be tainted by an infusion of Gentile philosophy, and the successive stages of corruption that afterwards took place: and this is an excellence peculiar to his history, and not to be found in any other that has as yet appeared in our language. He gives an account of the state of Unitarians at different periods; he takes notice of the early synods and the topics of discussion that took place in them; he marks the growth of heresies, the state of the Gnostics, Meletians, Donatists and Manichæans: but above all, he is particular in recording the dreadful persecutions and long-continued sufferings to which the early Christians were exposed during a period of near three hundred years, when all the civil power of the Roman empire was exerted *in vain* to extinguish the divine seed of Christianity, and to eradicate that plant which the heavenly Father had planted. He bestows deserved encomiums on the perseverance and fortitude of the martyrs in general, who bore torments too horrid and lingering almost to be described or contemplated, with exemplary patience and meekness; though he blames some of them who rashly provoked and courted persecution, and discovered too much sullenness, obstinacy, and contempt of their adversaries. The horrors of the last persecution under Dioclesian, are very particularly described; and (p. 495, &c. vol. 1.) some observations are introduced on this great persecution and the *effects* of it, which do honour to the pen of Dr. Priestley, and demonstrate, in opposition to Mr. Gibbon and all unbelievers, that Christianity by its own natural evidence, and the constancy of those who suffered for it, had out-grown Heathenism, and established

blished itself in the time of Constantine; not by external power or violence, or the mere authority and power of that Emperor, but by a general change of sentiment in its favour, arising from causes which had been long operating throughout the whole extent of the Roman empire.

Our Author gives an account of the constitution of the Christian church before the time of Constantine, of the edicts of that Emperor in favour of Christianity, of the Arian controversy and council of Nice, of circumstances relating to Constantine's conversion to Christianity, and his death, of the councils of Sardica, Ariminum and Seleucia, and other events that took place in the reign of Constantius, of Julian's prejudices against Christianity, his artifices to subvert it, and more direct attempts to undermine and gradually to extirpate it; and of the state of things in the succeeding reigns of Jovian, Valens, Theodosius, and Honorius, down to the year 475.

We come now to record a mournful and melancholy event, and ever to be regretted, if any event that has taken place under the government, and by the permission of a wise and good God can be called mournful, or furnish matter for lasting regret,

viz.

viz. the Riot at Birmingham. Various causes contributed to bring on this catastrophe. Our Author's repeated exertions in the cause of Unitarianism, produced a great alarm in the minds of many of the clergy. His attachment to the Dissenters, and his opposition to the test and corporation acts, with his fixed and rooted aversion to the ecclesiastical constitution of the church of England, increased this alarm. The circulation of the Familiar Letters to the inhabitants of Birmingham, the Discourse delivered to the supporters of the New College at Hackney, and above all his Letters to Mr. Burke, occasioned by his Reflections on the Revolution in France, published in 1791, inflamed the minds of the clergy, and the state politicians connected with them, to desperation. In these Letters our Author had confuted, with much spirit and humour, Mr. Burke's vaunting, frantic, and pedantic declamation in favour of civil establishments in religion, as well as signified his approbation of the French revolution. The storm which had been gathering for some time, and clouding the religious and political horizon, broke forth at Birmingham on the 14th of July, 1791, in a scene of dismal and diabolical burning and devastation,

vastation, too well known to require to be particularly described. After our Author had with difficulty made his escape to London, he addressed a letter, published in the Morning Chronicle, to the inhabitants of Birmingham, remonstrating with them in a calm and christian manner, on the enormity of the crime they had committed. He next published the Discourse on the Forgiveness of Injuries, before noticed: and last of all, he addressed the nation at large, in a work consisting of two parts, entitled, *An Appeal to the Public, on the Subject of the Riots in Birmingham.* London, 1791-1792.

The first part of this appeal contains a spirited dedication to the people of England, a preface in which a list is given of twenty-two addresses transmitted to the Author, seven from France, and fifteen from England, some of which are inserted at the end. A narrative is given respecting the Author's conduct and situation at Birmingham, the state of parties, and the circumstances attending the riot. The rest of the work is divided into ten sections, containing reflections arising from the subject and suitable to it. Various papers relating to the riot, or occasioned by it, are published in the appendix.

In

In the second part of the Appeal, our Author defends the account he had given of the riot, and the circumstances attending it in the first part, produces additional information on the subject, and replies to the charges and accusations of Mr. Burn. He makes observations on the proceedings in the Courts of Judicature, and the approbation of the riot, and the extent of high church principles which were the cause of it, in other parts of England. Nineteen pieces concerning the subject, or corroborating what our Author has advanced upon it, are printed in the Appendix.

The same year our Author published Original Letters, by the Rev. John Westley, and his friends, illustrative of his early History, with other curious papers, communicated by the late Rev. S. Badcock. To which is prefixed, An Address to the Methodists. Birmingham, 1791.

These letters are pious and devotional, but will be chiefly interesting to those who are attached to the religious opinions of the Methodists. Our Author, in his Address, endeavours to enlighten their minds, and recommends to them a more rational theology than their own.

Soon

Soon after, our Author addressed *Letters to the Members of the New Jerusalem Church, formed by Baron Swedenborg*. Birmingham, 1791.

It appears from the description that Dr. Priestley gives of these disciples of Baron Swedenborg, that they are a kind of visionary and mystical Unitarians. Our Author gives a short account of the life of Baron Swedenborg, a list of his numerous writings, and after applauding his disciples for their rejection of the doctrine of the Trinity, and attachment to the divine Unity, he points out the defects of some parts of their religious system, its inconsistency with the scriptures, and requests their attention to his own more simple scheme of religion. He quotes occasionally some passages from Baron Swedenborg's writings, and gives in the Appendix three large extracts.

The following year our Author published, *Letters to a Young Man, occasioned by Mr. Wakefield's Essay on Public Worship; to which is added a Reply to Mr. Evanson's Objections to the Observance of the Lord's Day*. London, 1792.

In the preface to this piece, our Author vindicates his deceased friend, Dr. Price, from some harsh censures of Mr. Wakefield. He considers the nature

nature of social prayer, and shews, in opposition to Mr. Wakefield, that it is a dictate both of reason and scripture. He replies to Mr. Wakefield's objections from the practice of Christ and his apostles, and shews the expediency and use of public worship. In reply to Mr. Evanson, he produces passages from Ignatius, Justin Martyr, Tertullian, and other early writers, which prove, that it was the practice of the Christians of the second and third centuries, to assemble for public worship on the Lord's day, and to consider it as a festival to be kept in joyful memory of our Lord's resurrection: he afterwards defends his sentiments on this subject from Mr. Evanson's objections.

Our Author having been elected a member of the National Convention, and in other ways treated with peculiar marks of respect by the French nation at this time, though he wisely declined removing into that country, yet thought it became him to testify his regard for them, by suggesting some useful advice on subjects of high importance.

At an early period, therefore, of the year 1793, he published at London, *Letters to the Philosophers and Politicians of France, on the Subject of Religion.*

These

These Letters are six in number. In the first, he endeavours to remove the prejudices that the French philosophers might be apt to conceive at the very mention of the term religion. In the second, he concisely and clearly proves the being of a God. In the third, he treats of the attributes and providence of God. In the fourth, he considers the evidence of the miracles performed in attestation of the Jewish and Christian religion. In the fifth, he gives cautions against superficial reasoning on this subject, replies to objections, and some passages of late French writers. In the sixth Letter, our Author shews that there is no necessary connection between religion and civil government, and that as legislators they ought not to interfere in the concerns of the former, but leave it entirely to its own operation, without civil aid or restraint. The composition of these letters is manly and spirited, and a great deal of important sentiment is expressed in a short compass.

Soon after the publication of these Letters, the war broke out between Great Britain and France, and a Fast-day having been appointed by public authority, our Author, on the 19th of April, delivered a discourse at the Gravel-pit Meeting, in Hackney, from Psalm xlvi. 1. which was afterwards published at London.

In this discourse our Author, without entering into any political discussion, considers the subject in a religious point of view, inculcating upon his hearers such sentiments as his text suggested, and the nature and circumstances of things required, and pointing out the great and important changes that would probably soon take place in the state of the world. In the preface to this sermon, our Author replies to some aspersions thrown upon him by Mr. Burke, and re-publishes a letter which he had before addressed to him in the Morning Chronicle.

The same year our Author published, *Letters to a Young Man, Part II. occasioned by Mr. Evanson's Treatise on the Dissonance of the Four generally received Evangelists.* London, 1793. It is matter of equal surprise and regret, that a man of Mr. Evanson's learning, ability and good character, should adopt so strange a paradox, as to set aside three of the four evangelists and a great many of the epistles, and to confine the whole authentic books of the New Testament to the Gospel of Luke, Acts of the

Apostles,

Apostles, 1st and 2d Thessalonians, &c. and Revelation of John. Dr. Priestley's reply consists of twelve letters. In the first and second, he considers the nature of historical evidence, illustrated by that of the propagation of Christianity, and the authority of the four gospels in general. In the third letter, he treats of the preference given by Mr. Evanson to the Gospel of Luke. In the fourth, fifth, sixth, and seventh letters, he replies to various objections of Mr. Evanson against the Gospel of Matthew. In the eighth, ninth and tenth letters, he defends the Gospels of Mark and John, and the Epistle to the Romans. In the eleventh letter, he replies to Mr. Evanson's objections to the Epistles to the Ephesians, Colossians, Philippians, Titus and Philemon, but passes over the objections to the Epistle to the Hebrews, the Epistle of James, those of Peter and of John, as thinking them perhaps of no weight. In the twelfth letter, our Author considers Mr. Evanson's proceedings as arbitrary, in making the Gospel of Luke his standard by which to examine the other Gospels. The Appendix contains as follows: 1. Remarks on some Passages in Mr. Evanson's Letter to the Bishop of Worcester. 2. Of the date of

Luke's

Luke's Gospel. 3. Of the Identity of Luke and Silas.

Another Fast-day, on account of the war with the French Republic, having been appointed in 1794, our Author delivered to his congregation, and immediately published, a *memorable* discourse on that occasion, entitled, *The present State of Europe compared with antient Prophecies; a Sermon preached at the Gravel-pit Meeting in Hackney, Feb.* 28, 1794, *being the day appointed for a General Fast. With a Preface, containing the Author's Reasons for leaving England.* London, 1794.

No person possessed of humane and virtuous sentiments, or even any degree of common liberality, can read the Preface to this Discourse, consisting of twenty-six pages, without admitting that Dr. Priestley had *sufficient*, yea *super-abundant* reasons, for leaving England; or without execrating the illiberal abuse thrown upon our Author, the unmerited ill usage he sustained, and the shocking infatuation of the times. That Dr. Priestley could not live without danger and molestation in his own country, that he was compelled to seek an asylum under the free and happy government of United America, and that

that eventually (contrary to his wishes) he should receive a *grave* in that land,* is a truth that cannot be denied; but which will reflect indelible disgrace on the temper and spirit of the high-church party, and that of the abject state politicians of his time.

The text is happily chosen. Matth. iii. 2. *Repent ye, for the kingdom of heaven is at hand.* Dr. Priestley considers these words as affording a stronger motive to repentance at present, than at the time they were originally spoken, as the approach of the kingdom of God is much nearer than it was at that period. By a large induction of passages from the prophetical parts of scripture, he proves that the kingdom of heaven, in the proper and complete sense

---

* The Preface concludes as follows :....." I sincerely wish my countrymen all happiness ; and when the time for reflection (which my absence may accelerate) shall come, my countrymen, I am confident, will do me more justice. They will be convinced, that every suspicion they have been led to entertain to my disadvantage, has been ill-founded, and that I have even some claim to their gratitude and esteem. In this case, I shall look with satisfaction to the time when, if my life be prolonged, I may visit my friends in this country ; and perhaps I may, notwithstanding my removal for the present, *find a grave*, as I believe is naturally the wish of every man, in the *land* that gave me birth.

of

of the words, refers to the millennial state of the church, and the world, when all anti-christian power shall be annihilated, the Jewish nation restored to the divine favour, and the kingdoms of this world shall become the kingdoms of Jehovah and his Christ. He shews from the same prophecies, that very calamitous events will precede this glorious state of things, which will particularly affect those parts of Europe that were formerly parts of the Roman empire, or have been subject to the Papal power, or concurred in oppressing the Jews in that state of dispersion, or that may hereafter endeavour to prevent their return and settlement in their own land. He views the great prevalence of infidelity as a definite mark or sign of those times that are to precede the second coming of our Lord; Luke xviii. 8. *When the son of man cometh, shall he find faith in the earth?* He observes a little before, "That those great troubles, so frequently mentioned in the antient prophecies, are now commencing, I do own I strongly suspect, as I intimated the last time that I addressed you on this occasion; and the events of the last year have contributed to strengthen that suspicion; the storm, however, may still blow over for the

the present, and the great scene of calamity be reserved for some future time, though I cannot think it will be deferred long." From all these considerations, he strongly enforces the duties of repentance, faith, and dependence on Divine Providence. In the Appendix, he has inserted some very pertinent extracts from Dr. Hartley's Observations on Man, and a sermon which had some time before been delivered in the chapel of Trinity College, Cambridge.

A few weeks after, our Author took leave of his congregation at Hackney, in a discourse from Acts xx. 32. entitled, *The Use of Christianity, especially in difficult times; a Sermon delivered at the Gravel-pit Meeting in Hackney, March* 30, 1794. *Being the Author's Farewell Discourse to his Congregation.*

This discourse is pathetic and affecting. Our Author considers the slight and precarious tenure of all human enjoyments and connexions, the many disappointments he had experienced, the derangement of his plans, and unexpected changes in his situation. He consoles himself and his flock under these trying circumstances, with the views and hopes that Christianity affords, recommends patience, for-

bearance

bearance and forgiveness, and a constant attachment and adherence to the principles of Dissenters and Unitarians. He expresses his satisfaction in the choice of Mr. Belsham as his successor, and concludes by addressing a few words to the many strangers who were present, shewing the moral tendency and innocence of Unitarian doctrine, and exculpating himself and his friends from having, in any respect, favoured sedition, or given any just ground for the calumnies and aspersions of their adversaries. The Appendix contains our Author's letter of resignation, with the reply of the congregation at Hackney to it, and addresses from the young men and young women, who attended lectures on the subject of naral and revealed religion; from the Unitarian Society, subscribed by the Rev. Mr. Lindsey as chairman, and from the united congregations of Protestant Dissenters at Birmingham. These addresses testify the highest good will and esteem for our Author, lament the depraved and malignant spirit of the times which occasioned his departure from England, and express the warmest wishes for his future happiness and prosperity.

About

Dr. Priestley's last publication in England, was a valuable and important work, entitled, *Discourses on the Evidence of Revealed Religion.* London, 1794.

These Discourses are affectionately dedicated to the Rev. Thomas Belsham, Tutor in Divinity in the New College, Hackney. Though the subject of these had been discussed by our Author before in several excellent compositions, of which an account has been given, yet as it is of vast extent, prime importance, and capable of various methods of illustration, our Author's labour upon it cannot be considered as superfluous, but highly necessary, reasonable and proper, to counteract the alarming progress of infidelity, more especially when he was about to take his leave of his native country.

The subject of the first Discourse is, *The Importance of Religion to enlarge the Mind of Man.* Here our Author evinces, in a strain of powerful argument, enforced with animated language, that the belief of a God, a Providence, and an actual state of things, has a natural tendency to improve the human mind, extend its comprehension, and raise it to the highest pitch of elevation; to produce an habitual devotion, and the sublime virtues of patience, meekness, forbearance and forgiveness: that the meanest and most unlearned Christian, possessed of these ennobling views and useful virtues is, and must be, superior to the best informed unbeliever. He shews (p. 17, 18) the danger of rejecting Christianity, and the debasement of character that is generally attendant upon it.

In the second Discourse, that revelation is the only remedy for idolatry and superstition, is shewn from the state of the Heathen world, and the lapses that large bodies of Christians themselves have made unto these lamentable errors, by not attending to the light of divine truth. The signal supernatural attestations by which the Jewish revelation was authenticated, in the deliverance of the Israelites from Egyptian bondage, the delivery of the law from Mount Sinai, the travels of the Jews in the wilderness, and their wonderful settlement in the land of Canaan, in the time of Joshua, are detailed in the third, fourth, and fifth Discourses. The sixth Discourse, contains general observations on the divine mission of Moses; at the end of which, some large and apposite quotations are introduced from the book

of

of Deutronomy. The seventh treats of the miraculous events from the time of Joshua to the Babylonish captivity. The eighth respects the prophecies concerning the dispersion and restoration of the Jews, in which the most material passages occurring in the Pentateuch and Prophets concerning these astonishing events, are inserted.

In the ninth and tenth Discourses, a concise but highly credible account is given of the miracles of Jesus and those of his Apostles. The last of these concludes in these words :....." On this firm basis, my Christian brethren, stands our faith ; and surely it stands upon a rock. It only requires an unbiassed mind, and especially a freedom from those vicious dispositions and pursuits which chiefly indispose men to the duties enjoined by the gospel, to perceive its evidence, and embrace it with joy." The eleventh Discourse treats of the resurrection of Jesus, and the twelfth contains a view of revealed religion. These two last had formerly been published separately by our Author, and are here re-printed, as having a natural relation to the subject.

The Appendix contains, 1. The Preface to the Discourse on the Resurrection of Jesus. 2. An

Address

Address to the Jews prefixed to the same Discourse. 3. The Preface to the Discourse, containing a View of Revealed Religion. 4. Dr. Priestley's Correspondence with Mr. Gibbon, relating to the proposed Discussion of the Evidences of Christianity contained in Part 1st, of the general conclusion to the History of the Corruptions, &c. little to the *credit*, but very agreeable to the character of that Unbeliever.

We now find Dr. Priestley safely landed on the American shore, happily free from the unmerited abuse, malignant aspersions, and insidious machinations of his enemies; though not without some trials to exercise his patience, even in this land of civil and religious liberty. The first thing he did in the literary line, (after replying to some friendly congratulatory addresses) was to re-print his *Appeal*, *Familiar Illustration*, *General View*, *&c.* with his Fast-day Sermon in 1794, Farewell Discourse at Hackney, and *Letters to the Philosophers and Politicians of France*; to which are prefixed, *Observations on the Cause of the General Prevalence of Infidelity*, which *Observations*, as they were afterwards re-printed and enlarged by our Author, we shall not stop to give an account of it at present.

These

These re-publications were well calculated to give the Americans an idea of his general principles, and reasons for leaving his native country. To the American edition of the *Appeal* is prefixed a short, but judicious and suitable Preface, bearing date Philadelphia, June 30, 1794.

His next literary labour in this country was, *An Answer to Mr. Paine's Age of Reason, being a Continuation of Letters to the Philosophers and Politicians of France, on the Subject of Religion; and of the Letters to a Philosophical Unbeliever.*

This publication contains three additional Letters to the Philosophers, &c. of France, viz. Letters 7, 8 and 9, by mistake printed 6, 7, 8. The seventh treats of the best method of communicating moral Instruction to Man; the eighth of Historical Evidence, and the ninth, of the Evidence of a future state. These three Letters are a very proper Sequel to the six former ones, and it is hoped may have had some good effect upon some individuals in France, if not upon the nation in general or its rulers.

The Letters to a Philosophical Unbeliever, Part III. are seven in number, and the subjects of them are as follows: 1. Of the Sufficiency of the Light of of Nature, for the Purpose of moral Instruction. 2. Of the Nature of Revelation, and its proper Evidence. 3. Of the Object of Christianity, and of the History of Jesus. 4. Of the proper Origin of the Scheme of Christianity, and Antiquity of the Books of the New Testament. 5. Of Mr. Paine's Ideas of the Doctrines and Principles of Christianity. 6. Of Prophecy. 7. The Conclusion. Under these seven general heads, Dr. Priestley has sufficiently confuted the first part of Mr. Paine's Age of Reason, (the second part was not then published) and exposed his quibbling arguments, ignorance, unacquaintedness with the style of the scriptures, and the customs of antiquity. Upon the whole, from a want of the necessary qualifications, and above all from a want of devotional sentiment, Mr. Paine was no more qualified to judge of the value and merit of the scriptures, or the proper evidence of revelation, than a blind man is qualified to judge of colours, or a deaf man of sounds. This piece was re-printed at London in 1795, with a large Preface of 37 pages, by Mr. Lindsey, expressive of his esteem for Dr. Priestley, giving an account of his situation in America, and defending

fending him from the calumnies and aspersions of his adversaries.

In 1796 and 1797, our Author delivered in Philadelphia, to very considerable audiences, a number of discourses in defence of divine revelation, which he afterwards published in 2 vols. 8vo. entitled, *Discourses relating to the Evidences of Revealed Religion, delivered in the Church of the Universalists, Philadelphia, and* (Vol. I.) *published at the request of many of the Hearers.* Philadelphia, 1796-97.

The first volume of these Discourses is dedicated to John Adams, Vice-President of the United States of America, betwixt whom and our Author a sincere friendship at that time subsisted, and who had been one of his constant hearers. The Preface contains some curious extracts from the third volume of *Asiatic Antiquities*, one of which relates to the Afgans, a people in the East-Indies, whom Sir William Jones supposes to be the descendants of the ten tribes carried into captivity by the Assyrians. These Discourses may be considered as a continuation of, or sequel to those published at London when our Author left England, and are in themselves highly valuable, and were calculated to have a very good

effect

effect in a country that had begun to be tainted with the infection of infidelity. They exhibit the following subjects, Vol. I. Serm. 1. The Importance of Religion. Serm. 2. Of the superior value of Revealed Religion. Serm. 3, 4. A View of Heathen Worship. Serm. 5, 6. The Excellence of the Mosaic Institutions. Serm. 7, 8. The Principles of the Heathen Philosophy compared with those of Revelation. Serm. 9, 10. The Evidence of the Mosaic and Christian Religions. Serm. 11. The Proofs of Revealed Religion from Prophecy. Serm. 12. Internal Evidence of Jesus being no Impostor. Serm. 13. The moral Influence of Christian Principles.

Vol. 2.

Serm. 1. (in two parts) The moral Design of Revelation. Serm. 2. (in four parts) Of the Authority assumed by Jesus, and the Dignity with which he spoke and acted. Serm. 3. (in two parts) The Doctrine of Jesus respecting Morals. Serm. 4. (in two parts) The Doctrine of a Resurrection, as taught by Jesus. Serm. 5. (in six parts) Of the Principles and Evidences of Mahometanism compared with those of Christianity. Serm. 6. (in two parts) The Genuineness of the Book of Daniel, and his prophetic

phetic character, vindicated. Serm. 7. (in two parts) Of the Prophecies concerning Antichrist. Appendix I. A Note concerning the figurative language of the Prophets and the sacred Writers. 2. Of the Influence of Mahometanism.

The same year in which the last volume of these Discourses appeared, our Author published, *Observations on the Increase of Infidelity, 3d edition. To which are added, Animadversions on the Writings of several modern Unbelievers, and especially the Ruins of Mr. Volney.* These Observations had made their appearance before in a more contracted form, and connected with other publications. They are here considerably enlarged, and contain many just and striking thoughts on the state of mind and habits of life, which either lead men to infidelity, or preserve them from it. Various quotations are introduced from the correspondence between Voltaire and D'Alembert, in order to exemplify the spirit and moral influence of infidelity; which do little credit to the cause itself, or its two famous champions and supporters. Some passages are quoted, p. 132, from Mr. Volney's *Ruins*, which savours of Atheism, or a very sceptical turn of thinking, and are justly animadverted upon by our Author. Excellent cautions and advices are given to rational Christians to conduct themselves aright during the present prevalence of infidelity, to view the subject in a serious and proper light, to set a just value on their religion, to derive comfort from it, and act according to it. In the Appendix, Mr. Volney's assertions concerning the theology of the ancient Egyptians, and the antiquity of the world, is examined, and Mr. Freret's account of the condition of the primitive Christians is considered.

In consequence of the notice taken of the writings of Mr. Volney in the preceding work, that writer, who was then residing in America, addressed a letter to our Author, which, by his account of it, appears to have been written with a considerable degree of peevishness and ill-humour, if not bad breeding. There was nothing so particularly harsh, I apprehend, in Dr. Priestley's censures, as to require a style of this kind. But infidels are a *genus irritabile*, as well as poets, and though liberal enough themselves in sarcasm and reproach, are angry if they meet with any in return. Our Author replied to him in a smart little pamphlet, entitled, *Letters to Mr. Volney,*

ney; occasioned by a work of his, entitled, Ruins, and his Letter to the Author. Philadelphia, 1797. In the first letter, our Author repels the censures of Mr. Volney. In the second, he shews the pernicious tendency of infidelity, and the bad effects that Mr. Volney's book, though destitute of any thing approaching to solid argument, yet recommended by the splendour of his imagination and fascinating charms of his diction, may have upon the minds of young and unthinking persons, by inducing them to reject the belief of a God, a Providence, and a future state, and to follow their prevailing inclination, whatever it be, under the pretence of following their only professed guide, nature. The third and fourth letters treat of the being of a God, and the evidences of revelation: and in the fifth letter, he reduces the several articles in dispute to the form of queries, eleven in number, which he proposes to Mr. Volney for his solution.

When our Author had delivered, in Philadelphia, the series of Discourses contained in his first volume to a mixed assembly, he thought it became him, as a man of sincerity and candour, to give some account of his own particular tenets, and to advise those

those who concurred with him in opinion, to form themselves into religious societies for divine worship and mutual edification. To effect this purpose, he preached and published a sermon, entitled, *Unitarianism explained and defended, in a Discourse delivered in the Church of the Universalists at Philadelphia*, 1796. Philadelphia, 1796.

In the Preface, he gives an address to the congregation, which he delivered after he had concluded his Discourses before mentioned, and proper advice to such conscientious Unitarians as cannot join in Trinitarian worship. His text is from Acts xvii. 18-20. At his first landing in America, having been excluded from every pulpit (except Princeton, where he was desired to preach) he considers his situation resembling very much that of Paul at Athens, described in his text. After reciting those points of religion which are of the highest importance, and held by all Christians, he mentions that there are other religious truths, though not of primary, yet of secondary, and of considerable importance, on which, for various reasons, he has thought it his duty not to be silent, especially in an age abounding with unbelievers. He observes, p. 7, that "Christianity,

tianity, besides being proved to be true, and indeed, as a necessary step in the proof of its truth, must be shewn to be *rational*, such as men of good sense can receive without abandoning the use of their reason, or making a sacrifice of it to what is called *faith*. The Author of our religion required no such sacrifice. He required of his disciples, that they should both *hear and understand* (Mark vii. 14.) what he delivered, which implies that he taught nothing that they were not capable of understanding, and which it was not their duty to endeavour to understand." He enlarges more particularly on the idolatrous *worship* of Jesus Christ, as God equal to the Father, the doctrine of the Trinity connected with it, and that of atonement, as the chief and most signal corruptions of Christianity, and the most obstinately retained; though he takes notice of the doctrines of predestination and original sin. He proposes and enforces the scriptural arguments by which the personal unity of God, the placability of his nature, and the proper humanity of Christ, are supported, and concludes with giving his assent in the fullest manner to the opinion of the final happiness of all the human race, maintained by the minister and congregation in

whose

whose place of worship he delivered this discourse.

An Unitarian society having been formed at Philadelphia, on the plan recommended by our Author, he published in the following year a small pamphlet, entitled, *An Address to the Unitarian Congregation at Philadelphia, delivered on Sunday, March* 5, 1797. Philadelphia, 1797.

In this little, animated, affectionate piece, our Author expresses his great satisfaction at the conduct of the members of this congregation, who, without waiting for the concurrence of the great, the wealthy, or the learned, or even that of any considerable number of persons of any class, had formed themselves into a society professedly *Unitarian*, in a part of the world in which no such thing existed before. He congratulates them on their freedom from penal laws, and that in this country the denying of the doctrine of the Trinity is not deemed to be *blasphemy*, punishable with confiscation of goods and imprisonment, as in England. While he inculcates upon them a just zeal for their own peculiar principles, and a steady adherence to them, he recommends a still greater attachment to the common principles of Christianity. He exhorts them to re-

spect

spect all Christians as such, and to be ready to inform them in a modest and respectful manner. He exposes the superstition of those who think that ministers, *regularly ordained*, are indispensably necessary to the constitution of a religious society, or the administration of Christian ordinances. He recommends to his Unitarian brethren the greatest purity of character, a constant attendance on public worship on the Lord's day, a strict care in the instruction of their children, and to forbear entangling themselves in the political concerns of this country. What a pity, that a religious society so formed, and having such an instructor, should have been of short duration. The yellow fever is said to have diminished their number and scattered them: but surely there was a *remnant left*, whose duty it was to support divine truth, and keep up all the ordinary forms of public worship, without which no sect can be expected to prosper, or their tenets make any considerable progress.

The same year our Author published, *An Outline of the Evidences of Revealed Religion*. Philadelphia, 1797.

The

The use and intention of this valuable little Tract, is well expressed in the two first sentences of the Preface. " When any controversy becomes very extensive, and of course complicated, branching itself out into many parts, the connection of which is not easily perceived, it is of great use to have a general outline of the whole; shewing the mutual relation of the parts, and their respective importance. This I have here endeavoured to do with regard to the evidences of divine revelation." This Tract is divided into six sections. 1. Of the Nature of Evidence, as applicable to this Inquiry. 2. Revelation not antecedently improbable. 3. The external evidence of Divine Revelation. 4. The Evidence of the Resurrection of Jesus. 5. The internal Evidence of the Jewish and Christian Revelations. 6. Various Objections to the Jewish and Christian Revelations considered.

Our Author also published, during his residence in Philadelphia this year, *The Case of poor Emigrants recommended, in a Discourse delivered at the University Hall in Phildadelphia, on Sunday, February* 19, 1797. Philadelphia, 1797.

Every

Every topic that could well be devised to awaken sympathy, or excite liberality, is here employed by our Author, in favour of emigrants that are in destitute circumstances, and stand in need of relief. He particularly considers the cases and situations of emigrants from Great Britain and Ireland; and reminds the Americans, that if not themselves, yet most certainly their ancestors, were strangers as well as they. Nor does he omit to take notice of the state of emigrants from France, the West Indies, and other countries, but recommends them all to the charitable assistance of their fellow creatures, whatever their political or religious principles may have been.

During the course of the year 1798, I find nothing published by our Author on the subject of theology; but as he did not know what it was to be idle, and never withdrew his attention from serious and important matters; he was then employed in collecting and arranging materials for a learned, ingenious, and elaborate work, which made its appearance in the following year, entitled, *A Comparison of the Institutions of Moses with those of the Hindoos and other ancient Nations; with Remarks on Mr. Dupuis's Origin of all Religions, the Laws and Institutions*

stitutions of Moses methodized, and an Address to the Jews on the present state of the World and the Prophecies relating to it. Northumberland, 1799.

This work is respectfully dedicated to the Duke of Grafton. The Author, in the Preface, makes honourable mention of Dr. Andrew Ross, who, he says, chiefly furnished him with the materials of his work, and shewed much zeal in promoting it. He gives a list of the titles of some of the books quoted in the work, and proposes a plan for a continually improving translation of the scriptures, with rules of translating.

This publication, so far as respects the Hindoos, is divided into twenty-four sections, with an Introduction. The contents of these sections are as follow: 1. Of the Antiquity of the Hindoo Nation and Religion. 2. Points of Resemblance between the Religion of the Hindoos and that of the Egyptians, Greeks, and other western nations. 3. Of the Vedas and other sacred books of the Hindoos. 4. Of the Agreement of the Hindoo Principles and Traditions, and those of other ancient Nations, with the writings of Moses. 5. Of the Creation, and the general Principles of the Hindoo Philosophy. 6. Of the Hindoo

Hindoo Polytheism and Idolatry. 7. Of the Religion of Egypt. 8. Of the Religion of the Schamans. 9. Of the different Casts among the Hindoos. 10. Of the Bramins. 11. Of the Prerogatives of the Mings. 12. Of the Situation of Women among the Hindoos. 13. Of the Devotion of the Hindoos. 14. Of the Restrictions of the Hindoos and other ancient Nations with respect to Food. 15. Of the Austerities of the Hindoos and others Heathen Nations. 16. Of the Hindoo Penances. 17. Of the Superstition of the Hindoos and others for the Cow, and also for the Elements of Fire and Water. 18. Of the licentious Rites of the Hindoo and other ancient Religions. 19. Of Charms and fortunate Times. 20. Of Trial by Ordeal. 21. Of various kinds of Superstition. 22. Of the Devotion of the modern Jews. 23. Of the Hindoo Doctrine of a future state. 24. Concluding Reflections.

Many strange and curious particulars are detailed under these different heads. The sections that appear to be the most interesting are, the 3d, 4th, 5th, 6th, 8th, 10th, 13th, 22d and 23d. The masterly observations of our Author interspersed in the work, and particularly the *concluding reflections*, must

must satisfy every inquisitive and truly impartial reader, that the institutions of this nation, so celebrated for its antiquity and extolled by sceptical writers, as well as those of all other Heathen nations, on a fair and just comparison, fall beyond measure short of the excellence, purity, and simplicity of the Mosaic doctrines and institutions; authenticated by a long train of stupenduous miracles performed in the presence of multitudes, confirmed by prophecies continually fulfilling, and worthy of the God from whence they derive their origin.

Our Author, in four sections, confutes the absurd paradoxes of Mr. Dupuis; and in the Appendix, the whimsical allegories of Mr. Boulanger, and gives an useful scheme of the Mosaic laws and institutions in fifteen parts, with references to the scriptures, to which a proper introduction is prefixed. The whole concludes with a solemn, affecting address to the Jews on their approaching glorious restoration, with a detail of some remarkable prophecies concerning it; and the mournful but happy change that will take place in their sentiments respecting Jesus and his religion, when this astonishing but certain event shall be accomplished.

Our

Our Author's next publication was, *An Inquiry into the Knowledge of the antient Hebrews concerning a future State.* London, 1801.

The manuscript of this piece had been sent over to England, and was published by the Rev. Mr. Lindsey, with a Preface, in which an extract is inserted of a letter from our Author, giving an account of his situation in America, accompanied with some general remarks on the subject of the Pamphlet. It is well known to those who are acquainted with the state of theological controversy in England, in the last century, that Bishop Warburton, the Author of the Divine Legation of Moses, has endeavoured to form an argument for the divinity of that legation, because the doctrine of immortality was not urged upon the Jews as the sanction of their ritual; and that different learned men, such as Doctors Sykes, Jortin, Stebbing, Hodge, &c. have adopted and defended opposite opinions with respect to the ancient Hebrews having, or not having had, a knowledge and expectation of a future state; and that some of these writers have been apt to apply critical violence to some passages of scripture, in order to adapt them to their favourite theories.

I have

I have met with no piece on the subject, that gives a clearer, more concise and candid account of this controversy, or which offers a better defence of the opinion of those who think that the ancient Jews had a real and certain knowledge of a future life, than a Treatise of Mr. Stephen Addington's, entitled, *A Dissertation on the Religious Knowledge of the antient Jews and Patriarchs; containing an Inquiry into the Evidence of their Belief, and Expectation of a future State*, 4to. London, 1757. This Treatise, however, though very full and complete, cannot be supposed to supersede our Author's useful labours on the same subject; besides, at this distance of time, it is probably very scarce, and not so well known as it deserves to be.

In discussing the subject before mentioned, Dr. Priestley pursues the following plan in five sections, there being no third section. 1. Presumptive arguments in favour of the antient Hebrews having the knowledge of a future state. 2. Of the allusions to a future judgment in the books of the Old Testament. 4. Of the belief of the antient Hebrews in a Resurrection. 5. Of the doctrine of the book of Job. 6. Of the fate of the wicked at the Resurrection.

tion. The passages from scripture, and the apocryphal books respecting the point in hand, are carefully collected, and arranged with judgment and propriety; and there is a pertinent quotation introduced from Josephus. A small piece is added to this work, called, *An Attempt to explain the Eighteenth Chapter of Isaiah.* About the time this *Attempt* was written, a gentleman in England, of some rank, had explained this chapter as respecting the French invasion of Egypt under Bonaparte, and Bishop Horsley, in a learned and critical Dissertation on the chapter, had endeavoured to confute his interpretation. For once we find Dr. Priestley and Bishop Horsley nearly agreed in opinion. Our Author considers the chapter as having no particular relation to Egypt, but as a prophetical denunciation of the judgments of God upon the nations, even the most distant, who had concurred in oppressing the Israelites, and a declaration of their future happy return.

The following year our Author published a Tract, called, *A Letter to an Antipædobaptist.* Northumberland, 1802.

In this piece, Dr. Priestley does not enter into the scripture doctrine on the subject, which, he says, has

has been so often discussed, that nothing now can be well urged with respect to it; but confines himself to the evidence of what was the doctrine and practice of the primitive Christians, those who lived nearest to the time of the Apostles. He alleges presumptive evidence and more direct arguments from the writings of the Fathers in favour of the practice of infant baptism, answers objections to these arguments, relates the origin of Antipædobaptism, and treats of dipping or sprinkling, and of the obligation of the rite of baptism itself. He bears particularly hard upon the late Mr. Robinson, of Cambridge. Some of the quotations our Author has produced, appear to have peculiar force to prove, that infant baptism was a rite generally practised, and thought necessary by the primitive Christians. The controversy, however, is of great extent, and has been largely and fully discussed by Gale, Wall, Gill, Robinson, and many other writers, who have embraced different sides of the question.

In the same and following year, our Author published the continuation of his Ecclesiastical History, which he had been employed in composing or revising at intervals for some years before, entitled, *A General*

General History of the Christian Church, from the Fall of the Western Empire to the present time. 4 vols. 8vo. Northumberland, 1802-1803.

In the second edition of the two first volumes of this history, which were re-printed at Northumberland after the Author's death, an additional section is inserted in the eighth period, relating to the Pretensions to Miracles, which began to be advanced and to gain credit in the course of the fourth century. In this section, the whimsical and ridiculous miracles of Anthony, as recorded in his life written by Athanasius; and those of Gregory of Neocæsarea, and of Martin of Pannonia, as related by other writers, are particularly detailed as specimens of the credulity and delusions of the times. There are also other additions and improvements, consisting of paragraphs inserted in different parts of the work.

With respect to the four volumes of the continuation, though in my apprehension they in some places fall short of the former part, in vigour of imagination, fertility of sentiment, and sprightliness of style, yet the work in general is of great value and utility. Apart from the consideration that the Author was now far advanced in years, and had encountered the rugged storms of adversity, the subject itself in great part did not admit of an equal display of genius. The mystical theology, and intricate and often trifling disputation of the middle ages, when general darkness had overspread the Christian world, and cramped and enervated the power of the human mind, were ill calculated to give that elevation to a writer, which the splendid scenes that took place in the commencement and progress of Christianity naturally inspired. Dr. Priestley, however, has made the most of his subject, and with exemplary candour bestows commendation upon whatever appearances of piety, ability, and useful learning the dark ages could supply him with. The work is dedicated to Thomas Jefferson, President of the United States, in terms of high respect and esteem, with a deserved encomium on the merit of that great man, and expressing the Author's satisfaction in spending the last years of his life under his just and equal administration. The Preface is worthy of a christian and a philosopher, abounding in solid and masterly reflections arising from the subject of his history. He views the long continued errors and prejudices of the

Christian

Christian world with an eye of compassion, and appears willing to make the best apology he could for them; and considers the circumstance of christianity working itself clear from its corruptions, and returning gradually to its primitive purity and excellence, as an unequivocal mark of its divine origin. He gives an account of the writers who furnished him with the materials of his history, and a list of the titles of such books as are quoted by the names only of the writers.

The first volume contains three periods, viz. from period 14. to 17. inclusive, extending from A. D. 475 to A. D. 1099; and comprehending, besides less considerable articles and events, the history of Eutychianism and Arianism, that of the Monks, the state of Heathenism, Judaism, and Christian sectaries, the progress and propagation of Christianity, the advancing power of the Popes, the controversy occasioned by the Monothelites, that relating to the worship of Images, and concerning the sense in which Christ is the Son of God, the rise and progress of Mahometanism, the power of Bishops and state of the Clergy, the disorders occasioned

sioned by the ignorance, superstition, and rapacity of the times, the intercourse between the Greek and Latin churches on the subject of the patriarch Photius, which led to their final separation, the controversy concerning Predestination, various instances of the claims of the Popes to ecclesiastical and civil power, and of the opposition that was sometimes made to them, the character of the Clergy, and the history of the first Crusade.

The second volume includes three periods, viz. from 18. to 20. inclusive, extending from the taking of Jerusalem by the crusaders in A. D. 1099, to the conclusion of the council of Constance, A. D. 1418. This volume contains the following particulars. The History of the Crusades continued, and their termination. Also, a continuation of the articles relating to the state of the Jews, Monks, Clergy, &c. The schisms in the church, and the transactions between the Popes and the Emperors of Germany. The History of Thomas Becket, Archbishop of Canterbury. Of the Paulicians, and other sectaries whose principles were similar, or bore some relation to those of the Manichæans. Of the Alligenses and Waldenses,

Waldenses, their opinions, the steps taken against them, and the persecutions they endured. Of Arnold of Brescia and the famous Abelard, with an account of his book on the subject of the Trinity, and the mortifying retractation he was obliged to make. Of the metaphysical speculations of Gilbert of Poree, Peter Lombard, and various opinions that were advanced in these times. Of the transactions of the Popes with the Emperors of Germany; Peter, King of Arragon; John, King of England, and their contests with Lewis of Bavaria. Of the difference between Pope Boniface VIII. with Philip le Bel King of France, and with the family of Colonna. Of the rise of the Franciscans and Dominicans, and some particulars relating to the Orders of Mendicants. Of the progress of the Inquisition, the state of heresy and Infidelity, and some particulars concerning the superstition and fabulous histories of the times. The history of the *great schism* which took place in the Popedom on the death of Gregory XI. An account of the Military Orders, and the suppression of the Knights Templars. Of the Fratricelli or Spiritual Franciscans. Of the Reformers prior to Wickliffe.

An

An account of Wickliffe, his tenets, followers, and the martyrdom of Lord Cobham, and others in England, for embracing and defending his opinions. Of John Hus and Jerome of Prague, their sufferings and death, and the proceedings of the council of Constance respecting them. Of various opinions theological and moral, held by Thomas Aquinas, Duns Scotus, and others, that were the subject of discussion in these times. Of the intercourse between the Greek and Latin churches, and attempts to procure an union between them. Our Author gives occasionally an account of the state of literature and learned men; and p. 155, has inserted a poem in Latin rhyme, written by Hildebert Bishop of Mans, entitled, *Oratio ad Dominum.*

In the third volume, the subject becomes more interesting, a long night of darkness, delusion and superstition, was now *far spent*, and the dawn of a great and necessary reformation began to appear. This volume contains two periods, viz. the 21st, extending from the conclusion of the council of Constance, in A. D. 1418, to the Reformation, A. D. 1517; and the 22d, from the beginning of the Refor-

mation

mation in Germany, A. D. 1517, to the conclusion of the council of Trent, in A. D. 1563. Under the 21st period, an account is given of the power of the Popes at this time, and the opposition that was made to it, of the transactions at the councils of Basil and Florence, Pisa and Lateran, of the Pragmatic Sanction, and Concordat of the Hussites, the long and cruel war they carried on, their defeat, and the rise of the *Unitas Fratrum*, or United Brethren, commonly called Moravians, of the attempts to unite the Greek and Latin churches, of various opinions advanced at this period, some of which are whimsical, but generally salutary and unfriendly to Popery, of Jetzer at Berne, and the imposture contrived by the Franciscans, of the clergy and monks, and articles relating to church discipline; and of miscellaneous articles, concerning the Waldenses and Jews, the propagation of Christianity, the Moors in Spain, the Turks, Unbelievers, Superstition, the Art of Printing, and the Progress of Literature.

Under the 22d period, after a recital of the gross abuses of the Romish church, and the depraved and scandalous

scandalous manners of many of its clergy, an account is given of the celebrated Martin Luther, and his proceedings in promoting the reformation. He was first roused to inquiry and exertion by the sale of *indulgences* in Germany by Tetzel and his companions, who had the authority of Leo X. then Pope, for this infamous traffic. Luther remonstrated against this shocking abuse with energy and effect, and proceeding step by step in his inquiries, was led by a gradual process to discover some of the most capital errors of the church of Rome. He did not fail to impart these discoveries to the public at large; and though at first he had his doubts, hesitations, and made some concessions and partial retractions, became in a few years a bold and decided reformer. When he made his appearance before the Diet at Worms, he resolutely adhered to his principles, and continued for more than twenty years afterwards to speak, act, and write, in favour of the reformation, with great intrepidity; and died peaceably in 1546, in a truly christian and edifying manner. Our Author has copiously detailed these and many other particulars concerning Luther, and connected with

such

such circumstances as render them really interesting. He relates the various artifices, expedients, and denunciations employed by Leo and his successors in the Popedom, and the Emperor Charles the Fifth, to subvert Luther and his adherents, and crush the rising reformation, which by a concurrence of happy events proved abortive; the powerful and kind protection afforded Luther by Frederick, Elector of Saxony; the state meetings and public assemblies that were held in different cities of Germany, occasioned by the rapid spread of Luther's tenets, &c.; the rupture that had taken place in the Catholic church; the confessions exhibited by the Protestants at Augsburgh; the events of the war that followed some years after; and the establishment of a permanent toleration of Protestantism in Germany, at the Diet held at Ratisbon in A. D. 1559.

The names of Zuinglius, Calvin, Bucer, Melancthon, Carolstadt, Oecolampadius, and other champions of the reformation, are frequently mentioned, particularly the two first and Melancthon, whose merit in advancing it were very great; though Calvin stained his name much by being concerned in the prosecution and condemnation of Servetus.

An account is given of the reformation in Switzerland, the Low Countries, Spain, France, England, Scotland, Sweden, Denmark, Poland, Hungary and Transylvania, with a relation of the sufferings of some of the martyrs. The English martyrs are more particularly described than those of other nations. This period also contains a history of the council of Trent, the early Anabaptists in Germany, and their disorderly proceedings, the Waldenses, and Bohemian brethren, the Unitarians, among which occur the names of Lewis Hetzer, an Anabaptist, who appeared so early as the year 1524, composed a treatise against the Trinity, which was suppressed by Zuinglius, and was put to death at Constance, John Campanus, Claudius, the celebrated Servetus, whose writings and sufferings are particularly described, Andrew Dudith, Lœlius Socinus, Francis David, De Wit, latinized *Spiritus*, Modrevius, &c. and miscellaneous articles relating to the times.

From the above list of Unitarians (to which more names might be added) some of whom were writers and men of learning and capacity, it appears that the cause of the divine unity did not want able advocates

advocates at the earliest period of the reformation. It was not indeed to be expected from the nature of things, that the tenets of these men could spread over the Christian world in general at that time; when they were placed in disadvantageous situations, their writings suppressed or burnt, themselves seized and put to death or banished, and equally cried down by the Roman Catholics and the popular leaders of the reformation. But we find, that such of them as made their way to Poland and Transylvania, and enjoyed the benefit of toleration, did not fail to make an impression on the minds of persons both of upper and lower rank, and that they left behind them proselytes and successors distinguished by their learning and ability.

Our Author has prefixed a short Preface to the fourth and last volume, containing a few additional observations to the large Preface of the first, an account of Emanuel Swedenborg and his tenets, and a list of the titles of books quoted in it. This volume contains two periods, viz. the 23d, extending from the conclusion of the council of Trent in A. D. 1563, to the revocation of the edict of Nantes in A. D. 1685; and the 24th, from the revocation of the

the edict of Nantes, in A. D. 1685, to the present time A. D. 1802. Under the 23d period, an account is given of the Popes, and the general character and state of the Catholic church; of the missions for the propagation of religion in the East Indies, China, Japan, Abyssinia, &c. by the Catholics and Protestants; of the controversies in the church of Rome respecting the power of the Pope, and the tenets of the Jesuits, Jansenists, &c.; of the religious orders and miracles ascribed to St. Anthony; of the Eastern churches, viz. the Greeks, Georgians, Mingrelians, Russians, Monophysites or Jacobites, Armenians, Nestorians, and Maronites; of the Protestants in the Valteline, the Waldenses, and the Bohemian brethren; the shocking massacres of the first by the Catholics, and the general persecution of all; of the Lutherans, their principles, and forms of church government; of the reformed who embrace Calvin's doctrines and discipline, and the difference between the system of Calvin and that of Zuinglius; of the Anabaptists, their tenets and discipline; of the Unitarians in Poland and Transylvania, their first simple catechism or confession of their principles, publish-

ed

ed in 1754, * many of their excellent writings published in the Bibliotheca Fratrum Polonorum, containing the works of Socinus, Crellius, Slightingius, and Wolzogenius ; some particulars of the life of Faustus Socinus, his differences with Francis David ; Simon Budneius denied the miraculous conception and was deposed ; the troubles of the Unitarians, and their expulsion from Poland, the martyrdom of Bartholomew Legat and Edward Wightman, and the sufferings of other Unitarians in England ; Mr. Biddle, his piety, learning, persecution, and death in prison ; of the Protestants in France, the massacre of them at Paris and other cities, the war that followed, the edict of Nantes enacted in their favour by Henry the Fourth, their declining state, the revocation of the edict of Nantes, and the

* Many years afterwards the Unitarians published another Catechism at Racow, mentioned by our Author. This they improved in the following editions of it. The best and most perfect edition is that republished in 1680, with notes composed by, or extracted from their best writers. It is perhaps the only Catechism, or Confession of Faith, published by a public religious body of men, that attempts a proof of Christianity, or exhibits a system of doctrines in the form of argumentative demonstration.

dreadful persecution that ensued; of the state of the reformed in the Netherlands and the United States; the controversy between the Calvinists and Arminians; the condemnation of the tenets of the latter at the synod of Dort; the institution of a benevolent Christian society called Collegiants; some visionary opinions; a curious dialogue between an Inquisitor and an Unitarian; of the state of the Puritans in England, the hard treatment they suffered in the reign of Elizabeth, and the various persecutions they endured for their non-conformity to the church of England, and conscientious adherence to their principles in the following reigns; of the Quakers, their tenets, discipline, peculiar practices, and firmness in bearing persecution; George Fox, a shoemaker in the north of England, the first of them whose name is known; William Penn, of England; and Robert Barclay, of Urie, in Scotland, their most celebrated writers; the former an Unitarian, and a strenuous opposer of the doctrine of atonement, and in company with his followers, the founder of Pennsylvania; of the state of religion in Scotland, in the reign of the Stuarts, and the cruel persecution and occasional resistance of the Presbyterians in that

country,

country, in the time of Charles the Second; and of miscellaneous articles respecting literature, religion, infidelity, and Judaism.

The 24th period reaches to the present time, and contains the following particulars. The consequences of the revocation of the edict of Nantes; the sufferings of the Protestants, particularly their ministers, the inhuman cruelties exercised towards them; the war in the Cevennes, with the long continued resistance, gallantry, and military exploits of the Camisards. A detail of the lingering and protracted sufferings, and almost super-human patience of these excellent and distinguished martyrs, M. Marolles, Le Fevre, and P. Mauru, *of whom the world was not worthy*. General articles relating to the Roman Catholics; the superior character of the Roman Pontiffs at this time; various debates, discussions and innovations that took place in the Romish church; the measures taken by the National Assembly at the French revolution, in regard to the clergy, freedom of religion, and general toleration; the attachment of Gregoire, Bishop of Blois, to religion; the regulations and establishments of Bonaparte respecting religion; his constitutions a great improve-

ment

ment on the former established religion of France, in many respects superior to that of England; but both infinitely inferior to the system adopted in the United States of America. The suppression of the Jesuits, their expulsion from all countries, and forlorn situation. The state of religion in Poland; events in Great Britain; the act of toleration granted to Dissenters at the revolution; the penal law enacted against Unitarians; oppressive acts made against the Dissenters in the reign of Queen Anne; the progress of free inquiry and liberality; the censures of Mr. Whiston's writings by the convocation, prevented from proceeding against him by the Queen; Dr. Samuel Clarke did not leave the church, or resign his benefice as Mr. Wharton did, though his opinions approached to those of the latter; Mr. Pierce, and other eminent Dissenters, became converts to the Arian doctrine; Dr. Lardner, an Unitarian; the doctrine of the proper humanity of Christ made its way into the established church; the petitioning clergymen, the Rev. Mr. Lindsey one of them, resigns his benefice, and institutes an Unitarian congregation in London, with a reformed liturgy; the application of Dissenters for relief of their

grievances,

grievances, are exempted from subscription to any of the thirty-nine articles; the toleration granted to Roman Catholics; new, but unsuccessful applications of the Dissenters to Parliament, for the repeal of unequal and disgraceful laws; the refusal of Parliament, in 1792, to repeal the act of William and Mary against Unitarians; the Irish association and rebellion, and the union of Great-Britain and Ireland; the riots at Birmingham, and its consequences, briefly hinted at in a note; the Methodists; Mr. John Wesley's labours, tenets, and the discipline he appointed for his sect; he ordained bishops for the mission of North America, after the acknowledgment of its independence on England; the eastern churches, and the Lutherans and Moravians in Europe; a particular account of the proceedings and progress of the latter; the progress of infidelity; an account of the English Deists, Lord Herbert of Cherbury, Mr. Blount, Mr. Toland, Lord Shaftsbury, Mr. Collins, Mr. Woolston, Mr. Tindal, Dr. Morgan, the anonymous authors of *Christianity not founded on argument*, and *the Resurrection of Jesus considered*, Mr. Chubb, Lord Bolingbroke, Mr. Hume, Mr. Gibbon; and of the French unbelievers,

 Voltaire,

Voltaire, Freret, *Systeme de la Nature*, Volney, Dupuis, with proper and seasonable reflections; the state of religion in the United States of North America; the opinions and conduct of the first settlers, and the general harmony and good neighbourhood that prevails among all the numerous sects and parties at present, in consequence of the exclusion of any establishment of religion in the general constitution. It is noted, p. 375, "that Unitarianism has of late made great progress in Boston and its neighbourhood, without exciting any alarm, though it is regarded with abhorrence almost every where else." Miscellaneous articles relating to the Quakers, Jews, Batavian republic, state of literature; an account given of Mr. Emlyn, his prosecution, trial, imprisonment and sufferings in Ireland, for writing in defence of the Unity of God, with the inscription on his tomb; the conclusion, with interesting observations suitable to the history, respecting the corruption and renovation of genuine Christianity; the progress of Unitarianism among the Dissenters in general, and even of late among the Methodists, and on the continent of Europe; subscription to articles of faith, the temper of the martyrs, and the disinterested

rested conduct of some persons who have made great sacrifices to truth and integrity; a chronological table of events, and a view of the succession of the Popes and principal temporal Sovereigns, is added to this last volume. I leave this subject with observing, that the former part of this Church history, lately re-printed, with the continuation, forming together a complete and uniform work in six volumes, 8vo. is a truly valuable acquisition to the republic of letters, and deserves the perusal of all the friends of Christian literature.

While the last volume of the Church History was in the press, our Author having visited Philadelphia for the last time, published in that city a Tract, entitled, *Socrates and Jesus compared.* Philadelphia, 1803.

A contrast or comparison between Jesus and Socrates, had been attempted before by John James Rousseau, and by a pious and eminent dissenting divine, Dr. Toulmin, to whom the above mentioned Tract is dedicated: so that our Author was by no means singular or original in the thought. It may deserve however to be considered with what propriety such comparisons are made, and on what principle

ple they can be vindicated. Properly speaking, to a confirmed and decided Christian, no comparison can be proposed between the super-naturally illuminated Jesus, the ambassador of God's grace and mercy to sinful men, and Socrates: the disparity is so great between the man to whom the spirit was imparted without measure, and in whom the fulness of divine wisdom and divine power dwelt, who authenticated his celestial mission by miracles, and confirmed it anew by his well attested resurrection from the dead; and an uninspired Heathen, who had nothing to guide him but nature's light, scattered tradition, Grecian philosophy, with his own good sense and honest disposition; that no striking point of resemblance appears on which a legitimate comparison can be founded, and any attempt of that kind seems, at least, superfluous and nugatory.

But with respect to unbelievers, the case is different: in their calendar, Jesus and Socrates are of equal rank, and stand on the same ground as moralists, philosophers, and public instructors; and a fair comparison may take place between them. If it shall be found then, after an impartial scrutiny into, and examination of the pretensions, doctrines, sentiments,

ments, life and character of each, that a *superlative* preference must be given to Jesus in every respect, though he lived in a country far less improved by science, philosophy, and polite literature, than Socrates did, the consideration may well strike an unbeliever with astonishment, arrest his progress in scepticism, excite him to a serious, careful, and impartial investigation of the evidences of Christianity, which may terminate in a cordial assent to its truth, produce a life of active piety and virtue, and secure a blessed immortality as the happy consequence of all.

With this last view, I apprehend, Dr. Priestley instituted his comparison of Jesus and Socrates; and indeed this is sufficiently obvious from his manner of treating the subject, and particularly from the concluding inferences, p. 48, &c. Our Author has extracted his authorities, for the account he gives of Socrates, chiefly from the *Memorabilia* of Xenophon and Plato's writings. These eminent writers were the cotemporaries and disciples of Socrates, and furnish sufficient materials for the subject. This Tract is divided into nine sections, the contents of which are as follow: 1. Of the Polytheism and Idolatry of

Socrates.

Socrates. 2. His Sentiments concerning the Gods and their Providence. 3. Of the excellent moral character of Socrates. 4. The Imperfection of his Ideas concerning Piety and Virtue in general. 5. Of Socrates's Belief in a future state. 6. Of the Dæmon of Socrates. 7. Of the Character and teaching of Socrates compared with those of Jesus. 8. Of the different Objects of the Instructions of Socrates and of Jesus. 9. Inferences to be drawn from the Comparison of Socrates and Jesus.

The comparison as it may be supposed, with respect to God, his attributes, providence, and a future state, moral and devotional sentiment, purity of character and life, manner of teaching and instructing mankind, and the effects and consequences, turns out eminently to the advantage of Jesus. The scriptures are pertinently quoted and applied, and such observations occasionally introduced through the whole, and strongly enforced in the last section, as seem well calculated to impart joy to the true Christian, and raise uneasy sensations in the minds of obstinate unbelievers who have any remains of candour left. But it is to be lamented, tha in this giddy, sceptical age, serious argument in favour of religion

religion is little regarded, and without attention to the subject, no good can be done even by the most ingenious and useful performances.

The apparent good intentions of our Author in writing and publishing this Tract, did not exempt him from censure. The late Dr. Linn, of Philadelphia, a young Calvinistic divine of the Presbyterian denomination in that city, felt himself disposed to enter the lists with our Author, and gather laurels in the field of disputation. He possessed a poetical genius, and some good natural talents improved by education. Perhaps he thought he might gain some advantage over a distinguished champion now advanced in years, and that even a failure would bring him no great discredit. Perhaps, also, he might be prompted by a zeal for his own sect, and a desire to hinder the effect of Dr. Priestley's writings. But whatever were his motives, or his virtues, he certainly was deficient in candour. He perverts or mistakes the design of Dr. Priestley, and casts harsh and injurious reflections on the character of Socrates. On the appearance of Dr. Linn's pamphlet, Dr. Priestley published a reply, entitled, *A Letter to the Rev. John Blair Linn, &c. in defence of the Pamphlet,*

phlet, entitled, Socrates and Jesus compared. Northumberland, 1803.

In this letter, our Author explains his intentions in writing his former Pamphlet, which indeed were obvious before, vindicates the character of Socrates, and replies to Dr. Linn's objections. As Dr. Linn had asserted in his publication, p. 6, "that it was Dr. Priestley's ardent design to lower Jesus Christ from that infinite station to which he and a certain number of Christians to which he belonged, suppose him to be entitled," &c.; and in the same page had farther said, "You imagine Jesus to be less than God. I hold him to be God;" and in p. 30, "the most holy and eternal Jehovah," Dr. Priestley examines these assertions by the scriptures, and quotes several passages from the gospels, which clearly prove the contrary. In the conclusion, he expresses himself with peculiar energy and vehemence against the doctrine of the Trinity and that of atonement; being led to do so, perhaps, by the harshness with which Dr. Linn had charged the crime of idolatry upon Socrates, but more so, because he had been informed that some of his orthodox friends in England had imagined that he was returning to the faith in which

which he had been educated. He was therefore willing to leave what might be called his *dying testimony* to his faith in the proper unity and perfect placability of the God and Father of all, &c."......See p. 55.

Dr. Linn made his appearance in this controversy a second time in a pretty large pamphlet, composed in some places with great asperity, rudeness, and drollery. He endeavours to make good his charges against Socrates, and appears to lay great stress on the imperfection of Dr. Priestley's account of the *difference* between Socrates and Jesus, and censures harshly what he had advanced on the subject of the Dæmon. His defence of the orthodox notions of the divinity of Christ, and the doctrine of atonement, is exceedingly weak, and discovers great ignorance as well as bigotry. He crouds his pages with quotations from orthodox writers, as if the points in question were to be decided by authority rather than the scriptures. The passages of sacred writ which he does quote, he misapplies; and relies implicitly on the English version, without attending to the various readings and emendations proposed by learned men. If Dr. Linn had only used harsh language in that part of his pamphlet where he replies to Dr. Priest-

 ley's

ley's strong declarations against the doctrines of the Trinity and atonement, it might have been excused on the principles of retaliation and necessary zeal, for what appeared to him to be the truth; but nothing can excuse the spirit of virulence and contemptuous insult that runs through the whole composition.

Our Author was in a very weakly state when this second publication of Dr. Linn reached him, and engaged in a composition that he wished much to finish. He however immediately wrote, and published a reply, entitled, *A Second Letter to the Rev. John Blair Linn, &c. in reply to his Defence of the Doctrines of the Divinity of Christ and Atonement.* Northumberland, 1803.

In this reply, our Author briefly vindicates his statement of the difference between Socrates and Jesus, and his motives in drawing the comparison; produces a passage from Xenophon in favour of Socrates, and the good character and behaviour of those who were his chosen companions and familiar friends, and remarks, that none of those whom Dr. Linn mentions under that character, were present at his death. He declares, that he professed not to have

any

any fixed opinion with respect to the Dæmon of Socrates, and leaves the subject in the following words. "Whether Socrates was a little better, or a little worse, than he has been represented, is of little consequence to my object in writing, which I am sorry to find it is not in my power to make you understand." He sufficiently confutes what Dr. Linn has advanced in favour of the divinity of Christ and the doctrine of the Trinity, in those passages he touches upon, and refers to his small Tracts for others not particularly noticed. He states, in opposition to Dr. Linn, the rapid increase of Unitarians in England, and the congregations of that denomination in the eastern states of United America, with other facts. His observations on the doctrine of atonement are general, and he concludes with some account of the progress of his opinions, his conduct as a preacher and in controversy, and the extensive sale of some of his Unitarian Tracts. He takes notice in the beginning of this piece of Dr. Linn's rudeness and asperity, but treats him with much less severity than he deserved. It is remarkable, that this young violent controversialist did not survive Dr. Priestley

above

above six months, and died at Philadelphia in the 27th year of his age.

The same year our Author published a Tract, entitled, *The Originality and superior Excellence of the Mosaic Institutions demonstrated.* Northumberland, 1803.

This valuable Dissertation is properly an Appendix to his Notes on the five books of Moses, and contained in the first volume of the *Notes on all the Books of Scripture.* But our Author printed it separately, from the benevolent purpose of giving it a more general circulation. No sensible person, divested of prejudice, can read it and the Preface, without acknowledging the justness of the title, and inferring from the originality and superiority of the laws and institutions of Moses, to those of all the nations with which they are compared, as well as the peculiar circumstances of the Jewish people, that the claim of their great Legislator to a mission from Jehovah, the God of the whole earth, was just and well founded.

Prior to our Author's death, a considerable part of the *Notes* mentioned in the preceding article were printed, the remainder were put to the press, and the

whole

whole published by his son, Mr. Priestley, after that event, entitled, *Notes on all the Books of Scripture, for the Use of the Pulpit and private Families.* 4 vols. 8vo. Northumberland, 1803-4.

A pious and affectionate dedication is prefixed to this work, addressed to William Russel, Esq. and the other members of the congregation of Protestant Dissenters of the New Meeting at Birmingham.

In the Preface, abounding with useful and important matter, our Author informs his readers, that he made a considerable progress in this work when the riots at Birmingham took place, and destroyed a great part of what he had composed of these notes and transcribed for the press, and that having abundant leisure since his settlement in this country, he had re-composed those that were destroyed, and completed the rest in the best manner he could, being urged both by his own liking to the work, and the frequent requests of his friends in England. He modestly observes, p. viii. "Though I have spared no pains to make this work as perfect as I could, too much must not be expected from it, because my plan does not comprehend every thing. If *critics* and *scholars* look into it for the solution of all such difficulties

difficulties as they particularly wished to see discussed, they will be disappointed. These Notes will appear, from the account I have given of them, to have been composed for the use of *unlearned*, though liberal and intelligent Christians; for of such my congregations consisted. Nothing however, which such persons are much interested to know, I have passed without notice, whether I could explain the passages to my own satisfaction or not, and a few observations of a more critical nature I have added since; but which, if any minister, chuse to avail himself of my labour, he may omit or change, as he shall think proper. The same may be done by those masters of families, whose laudable custom it is to read portions of the scripture to their children and servants, and to those it is my wish more particularly to recommend what I have done."

From the account our Author here gives of the plan of his work, and his motives in composing it, it is obvious that he did not intend it so much for the use of the learned as that of liberal congregations, and intelligent private Christians. It is, however, doing no more than justice to say, that he has performed more than he promised. The scholar and

critic,

critic, in perusing these Notes, will be gratified by improvements suggested to the common version of the scriptures, and by curious particulars occasionally interspersed relating to ancient customs, usages and manners. Our Author has not only availed himself of the remarks of former commentators, but has inserted from the narratives of late travellers such hints and observations as were calculated to throw light upon obscure passages, and his own good sense and general knowledge of biblical literature, has frequently led him to make striking and original remarks.

It cannot be expected that I should enter into a minute detail of a work of this nature, consisting of so many detached particulars. I shall therefore confine myself to a few general observations. Our Author justly considers Moses as the writer of the Pentateuch, or the five first books of the Old Testament. He thinks it not improbable that the account of the five first days of the work of creation, might be communicated to Moses by revelation. He thinks that *days* may not be literally meant, but certain portions or periods of time; and that it is highly probable that the creation of animals took place at different

different periods, that of the carnivorous, for example, long after the world was stocked with those of the graminivorous kind. His account of the dispersion of mankind, after their attempt to build the tower of Babel, and of the nations to which the names recorded by Moses bear relation, is very exact and particular. His observations on the Patriarchs, and the Iraelites their descendants, and their history, laws, and institutions as related in the five books of Moses, are pertinent and ingenious. The Dissertation before noticed, is placed at the end of the notes on Deuteronomy. The other historical books of the Old Testament, are illustrated with equal ingenuity. The Notes on the books of Job, Psalms, Proverbs, and Ecclesiastes, are moral, sentimental, devotional, and occasionally critical. Our Author considers the book called Solomon's Song, as a poetical composition, having no mystical or spiritual meaning. In the prophetical books, he endeavours to illustrate the sublime figures and allusions of the writers, and never loses sight of what appears to have been one great object of these compositions, viz. the happy return of the Jews from their great and general dispersion, their acceptance

of

of Jesus as the Messiah, and the renovated state of the world after a period of great trouble and calamity. After the book of Ezekiel, a Dissertation is inserted, containing *General Observations on the Subject of Prophecy*. The historical prophecies of Daniel are also happily illustrated, but the last part of them can only admit of probable conjecture.

The Gospels are brought into the form of an harmony, but a table is placed at the end of the last volume to direct the place in which any passage may be found. The paraphrases on the discourses of Jesus, are taken from the Author's Harmony, before published in 4to. with many additional notes. In commenting on the first chapters of Matthew and Luke, our Author, with commendable delicacy, avoids explaining himself, particularly on the subject of the miraculous conception. The Notes on the Acts of the Apostles, as well as those on the Gospels, are very instructive and interesting.

Our Author's comments on the epistles of Paul and the other epistles, are judicious and practical, and the occasional paraphrases, by concentrating the sense and bringing it home to the heart, have a fine and edifying effect. On 2 Peter iii. 7, &c. he admits the possibility of the world being destroyed by fire, or any other means; but he adds, "the language of the apostle in this place is probably figurative, and only descriptive of those great changes in the state of the world which will precede the second coming of Christ, and the commencement of his proper kingdom. What follows, p. 544, &c. is well worth attending to.

Our Author bears a full and ample testimony to the authenticity of the Revelation. He does not differ materially from former commentators with respect to the seals, trumpets, and the different visions respecting the church, witnesses, and anti-christian powers represented by savage beasts. He considers Ch. xi. 12. as respecting the French revolution; and if so, and the last verses of this chapter be rightly placed, "that the sounding of the seventh trumpet will immediately follow the termination of the persecution of the witnesses, and the revolution which was co-incident with that event. See vol. 4, p. 509, &c. On Ch. xiv. 6, 7. where intimation is given by an angel of a purer state of the gospel, he observes, "These new preachers will probably be Unitarians, confining their worship to the one God, the

the maker of all things, and warning all people to keep themselves clear of every thing tending to idolatry, or any other worship than that of the God and the Father of Jesus Christ." He gives his conjectures concerning the vials, and intimates that the lamentation over Rome (Ch. xviii. 10, &c.) as if it were a commercial city resembling antient Tyre, may respect England. He thinks that Christ will make his personal appearance at the Millennium; that all who have suffered for his cause, and perhaps all good Christians, will reign with him, and assigns his reasons for thinking so at considerable length; but does not believe that all who do not share in the first resurrection will perish. He considers the invasion of Gog and Magog, mentioned by Ezekiel, and that by John in the Revelation, as relating to the very same persons and period of time; but different from the invasion described by Zechariah. He thinks that the expression, *being cast into a lake of fire*, Ch. xx. 15, as well as the literal sense of many passages of scripture, may denote the extinction or annihilation of wicked men, but that other reasons, which he states, may incline us to entertain the hope of the final restoration of the wicked by means

of

of a course of discipline in a future state. See p. 661.

The new heavens and new earth, Ch. xxi. I. "he thinks, can only mean a renewed and improved state of this earth, in consequence of which it will be so different from the present, as to deserve to be called a *new earth*:" but that pain, troubles and death, will be removed from this happy state. Through the whole work, our Author has been careful, where his subject led him to it, to enforce the arguments for the unity of God and proper humanity of Christ, as important and necessary points of Christian doctrine. On this, as well as on other accounts, these *Notes* appear to me to deserve the attention of intelligent Ministers and Christians in general, and to be extremely well adapted for the use of those Unitarian societies who are deprived of the advantage of a learned teacher. They are the only set of Notes on the *whole* bible, in our language, that can be properly called an Unitarian Commentary, and deserve to be classed with the Latin Annotations of Socinus, Crellius, Slightingius and Woltzogenius.

The last period of Dr. Priestley's life, and while he was in a state of great bodily weakness, was employed

ployed in composing a very important Treatise, published by his son after his death, entitled, *The Doctrines of Heathen Philosophy, compared with those of Revelation.* Northumberland, 1804.

This work is dedicated to the Rev. Joseph Berington, a Catholic priest in England, and to the Rev. William White, a bishop of the Episcopalian church in the United States. Our Author assigns a very handsome reason for this dedication. "Entertaining the highest respect for your characters, as men and as christians, I do it *because* we differ; to shew, with respect to a subject in which we are equally interested, as in that of this work, that I regard all that bear the Christian name, how widely distant soever their different churches and creeds may be, as friends and brethren, and therefore entitled, by the express directions of our common Saviour, to particular respect and attention as such."

The subjects of this Treatise, apart from the Tract, *Socrates and Jesus compared*, which is reprinted in it, are the following, viz. The state of religious and moral principles in Greece *before* the time of Pythagoras, consisting of an Introduction and six Sections. The Philosophy of Pythagoras, with

an

an Introduction and four Sections. The Philosophy of Aristotle, with an Introduction and three Sections. The stoical Philosophy of Marcus Antoninus and Epictetus, with an Introduction and three Sections. The Philosophy of Epicurus, with an Introduction and three Sections.

It is obvious, that this work is an extension of the plan and object our Author had in view in composing his Tract concerning Socrates, &c. He has selected with care, fidelity, and candour, the most pertinent passages from the Heathen poets and philosophers respecting all the topics included in these Dissertations. He has judiciously exemplified their turn of thinking, and appears willing to do them the most ample justice. The appeals to scripture, and the sentiments of the sacred writers, are not so numerous as those in Socrates, but they are sufficient for the purpose; and in the conclusion, a brief summary is given of the sentiments of the more intelligent Greeks and philosophers that succeeded them, in which the manifest superiority of the Gospel of Jesus Christ, and the permanent good effects it produced in enlightening and reforming the world, are shewn. From this work it appears, that

that Plato, Aristole, and other renowned sages of antiquity, were not so destitute of the knowledge of God, and of many points of moral duty, as of the right and consistent use and application of that knowledge. As the apostle Paul says, *when they knew God, they glorified him not as God, neither were thankful, but became vain in their imaginations, and their foolish heart was darkened. Professing themselves to be wise, they became fools, &c.* Plato, and indeed all the philosophers, recommend the observance of the idolatrous rites and ceremonies established by the laws of their country; and the former manifests the most intolerant sentiments against any who might attempt to institute a separate worship, and prescribes severe punishments to be inflicted upon them. Marcus Antoninus reduced this system to practice, and notwithstanding all his merit as a philosopher, was a cruel and unrelenting persecutor of Christians. The application and improvement of the whole subject may be made in the words of the apostle before quoted, 1 Cor. i. 21. *For after that, in the wisdom of God,* (by the display of the external phænomena of nature) *the world by wisdom,* (i. e. the exercise of reason or philosophy) *knew not God;*

God; (practically and effectually) *it pleased God by the foolishness of preaching*, (a doctrine of simplicity propagated chiefly by unlearned men) *to save them* that believe; that is, to effect a great and lasting reformation in the sentiments and practice of all who embraced the Gospel.

After our Author's death, there was printed at Philadelphia, a very useful composition of his, entitled, *Index to the Bible; in which the various subjects which occur in the Scriptures, are alphabetically arranged, with accurate References to all the Books of the Old and New Testaments, designed to facilitate the Study of those invaluable Records.* Philadelphia, 1804.

This publication is calculated to be of eminent service to those who have a relish for the scriptures, and who would wish to find readily the account of any fact, rite, ceremony, precept, &c. contained in those sacred books, without the trouble of much searching. A full account of the plan of it, and the Author's care in composing it, are given in the Preface.

Four Discourses, composed by our Author, were also published after his decease. The subjects of these

these are the following: 1. The Duty of Mutual Exhortation. 2. Faith and Patience. 3. The Change which took place in the character of the Apostles after the resurrection of Jesus Christ, in two parts. These Sermons are pious and practical, and cannot fail to have a favourable influence upon persons who read them with minds disposed to profit by the salutary admonitions contained in them.

Having now finished my review of our Author's theological publications, I shall close the subject with a few general reflections.

Dr. Priestley's choice of the Christian Ministry, and attachment to it as a profession, must have been founded upon motives of the purest kind, to have enabled him to support the difficulties and discouragements he had to encounter at his first appearance in public life. We have seen that he was far from being a popular preacher. Neither the principles he had espoused, the doctrines he taught, or his manner of address, were at all calculated to catch the giddy and unthinking, or even to recommend him to the esteem of serious and grave people who had embraced the Calvinistic system, at that time very prevalent among the Dissenters. But we do not find that he ever dissembled, or even concealed his principles in conversation with his hearers; though he did not think it necessary to insist much upon them from the pulpit. A mind less ardent and less disinterested than *his was*, could hardly have borne the inconveniencies of a narrow and insufficient income with patience and serenity; but would have hasted to relieve itself by embracing an employment more lucrative, independent, and respectable in the world's estimation.

His active penetrating genius led him early to examination and inquiry, and consequently to make some considerable changes of opinion with respect to those doctrines in the belief of which he had been educated, and which were in themselves absurd and unscriptural. But if he was not precipitate or overhasty in making these early changes, he was still less inclined to abandon those tenets which he had adopted after mature examination, and which had any appearance of reason or scriptural authority to recommend them. He continued long attached to Arianism, and notwithstanding his respect for Dr. Lardner and his esteem for his writings, he could not prevail upon himself to embrace the Socinian system during

ing the life-time of that writer, and did not become a convert to it, till after his decease.

His residence and intimacy with Lord Shelburne brought within his reach, and presented to his view, the prospect of political or ecclesiastical preferment; but he had the vittue and the fortitude to decline all connection with either, and continue a Dissenting Minister: and though he did not at that time statedly officiate in any congregation, he was always ready to afford his assistance to his brethren, and was as much engaged in theological studies and publications as at any other period of his life. Under two different administrations, overtures were made to him to accept a pension from government; but with a magnanimity peculiar to himself, and which has no example to countenance it in any other literary character of the age, he resisted the temptation, and preserved his independence. He accepted, however, assistance from distinguished and worthy private characters, and was by no means backward or reserved in expressing his gratitude for their donations.

No writer of the late century (Doctors Lardner and Leland excepted) wrote so much as he did in defence of Revelation, and under such a variety of

forms:

forms: he has placed the evidence of Judaism and Christianity almost under every point of view that could strike or affect the mind, and nearly exhausted the subject: he has reduced unbelievers to the *dilemma* of either embracing Christianity, or accounting for past and present appearances in a satisfactory manner, which it is impossible for them to do, and none of them have even attempted it.

Though he could not be called an Orator in the popular sense of the word, the Discourses he has published are by no means destitute of energy or pathos, or that kind of eloquence which is calculated to have a good effect on a sensible and delicate mind, and in general may be recommended as excellent models of composition for the pulpit. He usually gave short and useful expositions of some portion of scripture before he delivered his sermons in public, and these expositions laid the foundation of his *Notes on all the books of Scripture*, of which the public are now in possession.

But his labours as a Christian Minister were not merely confined to the pulpit: he made the religious instruction of youth an object of his particular care in the different congregations over which he presided,

presided, arranged them into distinct classes according to their age and sex, and with much ingenuity adapted his method of teaching to their different capacities. He lived on the most friendly footing with the congregations of Leeds, Birmingham, and Hackney, which he successively served; and received the most ample testimonies from each * in favour of the utility and fidelity of his ministerial labours in general, and particularly of their grateful sense of his assiduity and diligence in forming the minds of their children, and leading them to the knowledge as well as the practice of Christianity. His prayers were fine pieces of devotional composition, and had a considerable variety in them: these he committed to writing, and read, for the sake of greater distinctness and accuracy. He composed a variety of Catechisms for the improvement of youth, prayers for the

* Besides the ample testimonies of approbation which Dr. Priestley received from the three congregations above mentioned in his life-time, the congregation of the *New Meeting* at Birmingham have erected a monument to his memory in their place of worship since his decease, sufficiently expressive of his merit and their attachment; which will be found at the end of this work.

use of families, and devotional offices for that of Unitarian Societies.

The cause of civil and religious liberty is particularly indebted to his labours. He was closely and fervently attached to the credit and interests of the Protestant Dissenters, and stood forth as their champion and defender on different occasions; and surely his strenuous exertions, and various well-composed and spirited publications in their behalf, will not be forgotten by that respectable body of men-

The Unitarians can *never forget* his attachment to their cause, and the faithful and important services he performed by the publication of numerous works, and treatises, large and small, in their favour, and particularly in exploring the dark and intricate regions of ecclesiastical antiquity, in order more fully to corroborate their system; and maintaining the ground he had taken, and the advantages he had gained by superior research, perseverance, and acuteness.

When in the course of Providence he was called on to suffer persecution, obloquy and reproach, he supported these evils with exemplary fortitude and patience, and manifested a truly christian spirit of

candour

candour and forgiveness. When residing in America, and separated from his former congregations and religious friends, he still kept up the forms of public worship on the Lord's day, and neither the smallness of his auditory, nor the odium under which some of his tenets lay, could prevent him from discharging these labours of love.

Not only his numerous works in general, but even his prefaces and dedications, are pregnant with important matter and sentiment, and deserve to be read. He was indeed a most extraordinary man, and making candid allowances for some mistakes and inadvertencies to which all controversial writers are more or less liable, may be stiled the *Luminary* of his day. He retained the vigour of his faculties and his habits of unremitted exertion to the last; for in his latest compositions, there are no marks of intellectual decay, and he died with serenity and composure, after having finished the labours of a long and useful life.

CALEDONICUS AMERICANUS.

Northumberland, Pennsylvania, 1804,
10th July, 1805.

THIS TABLET
Is consecrated to the Memory of the
REV. JOSEPH PRIESTLEY, L. L. D.
by his affectionate Congregation,
in Testimony
of their Gratitude for his faithful Attention
to their spiritual Improvement,
and for his peculiar Diligence in training up their Youth
to rational Piety and genuine Virtue:
of their Respect for his great and various Talents,
which were uniformly directed to the noblest Purposes;
and of their Veneration
for the pure, benevolent, and holy Principles,
which through the trying Vicissitudes of Life,
and in the awful hour of Death,
animated him with the hope of a blessed Immortality.

His Discoveries as a Philosopher
will never cease to be remembered and admired
by the ablest Improvers of Science.
His Firmness as an Advocate of Liberty,
and his Sincerity as an Expounder of the Scriptures,
endeared him to many
of his enlightened and unprejudiced Contemporaries.
His Example as a Christian
will be instructive to the Wise, and interesting to the Good,
of every Country, and in every Age.
He was born at Fieldhead, near Leeds, in Yorkshire,
March 24, A. D. 1733.
Was chosen a Minister of this Chapel, Dec. 31, 1780.
Continued in that Office Ten Years and Six Months.
Embarked for America, April 7, 1794.
Died at Northumberland, in Pennsylvania, Feb. 6, 1804.

FOUR DISCOURSES

INTENDED TO HAVE BEEN DELIVERED

AT

PHILADELPHIA,

BY JOSEPH PRIESTLEY, L. L. D. F. R. S.

PUBLISHED BY DESIRE OF THE AUTHOR.

NORTHUMBERLAND:

PRINTED BY JOHN BINNS.

....

1806.

ON

THE DUTY

OF

MUTUAL EXHORTATION.

Exhort one another daily while it is called to day, lest any of you be hardened by the deceitfulness of sin. HEB. iii. 13.

THIS advice of the author of this epistle is not less seasonable at the present day than when it was given. It is even more deserving of attention now than it was then. At that time the christian church was in a state of persecution. At least the open profession of christianity was attended with more danger than it is at present. It was not then patronized by the great, the learned, or the fashionable; but was a *sect every where spoken against*, and the teachers of it were generally considered as *men who turned the world upside down*, the enemies of peace, and the authors of innovation and revolution.

Such, indeed, will ever be the character of *reformers.* It was so in every period of the reformation from popery. In this light were Wickliffe, Huss, Luther, Calvin, and Socinus considered in their

day;

day; and such is the light in which every person who in the present times, having by any means acquired more light than others, is desirous of communicating it, and to improve upon any established system, must expect to stand. The bulk of mankind wish to be at their ease, and not to have their opinions, any more than their property, or their government, disturbed. Being satisfied with their present situation, they naturally dislike any change, lest it should be for the worse. The situation of a reformer must, therefore, require great fortitude, the courage of the lion, as well as the wisdom of the serpent, and the innocence of the dove.

These virtues are equally necessary in our times, as far as they bear the same character; but they are only peculiarly requisite for reformers, and their immediate followers. With respect to christianity in general, the profession of it is not, at least in this country, at all disreputable. On the contrary, it is rather disreputable not to be a christian; and I rejoice that it is so, and that infidelity has not made so much progress as to make it otherwise. And I am willing to think that the seasonable and temperate answers which several learned christians have given to the numerous writings of ignorant and petulant unbelievers, have been a check, at least with all sober minded and thinking men, to the late alarming increase of infidelity.

But because the profession of christianity is not disreputable, is the genuine spirit of it more readi-

ly

ly imbibed, and the practice of its precepts more easy? By no means. There is another enemy to contend with, far more to be dreaded than open violence, against which it behoves us to be upon our guard, if we wish to have any thing more of christianity than the name, which alone will avail us nothing; and from the insidious and unsuspected attacks of the enemy, we have no means of escaping, as we might have from those of an open persecution.

This enemy is the world in which we live, and the intercourse we must have with it. For now, as much as ever, to be the *friend of what* may properly be called *the world*, is to be *the enemy of God. Love not the world* says the apostle John, *nor the things that are in the world. If any man love the world, the love of the father is not in him. For all that is in the world, the lust of the flesh, the lust of the eye, and the pride of life, is not of the Father, but is of the world. And the world passes away, and the lust thereof; but he that doth the will of God abideth for ever.*

In order to feel, and consequently to act, as becomes a christian, and this in an uniform and steady manner, the principles of christianity must be attended to, and never lost sight of. In time of persecution the distinction between christians and other persons who are not christians is constantly kept up. For then the mere profession of christianity makes men liable to suffering, and often to death; and when men are in danger of suffering for

any

any thing, as well as when they have the hope of gaining by any thing, they will give the closest attention to it. Their hopes or their fears cannot fail to keep their attention sufficiently awake.

When a man is willing to give up his property, and even his life, for the sake of any thing, he must set a high value upon it. He will cherish the thought of it, as what is dearer to him than any thing else. In such times, therefore, no man would for a moment forget that he was a christian. The precepts and maxims of christianity would be familiar to his mind, and have the greatest weight with him.

But this is not the case in such times as these in which we live. There is very little in a man's outward circumstances depending on his being a christian or no christian. The behaviour of other persons toward him hes no relation to that distinction; so that he has nothing either to hope or to fear from the consideration of it, there being nothing that necessarily forces, or that very loudly calls for, his attention to it. All the attention that, in these circumstances, he does give to it must be wholly voluntary, the spontaneous effort of his own mind. If his mind be much occupied by other things, he will necessarily relax in that attention, and if he intirely drop his attention to the principles of christianity; if all his thoughts, and all his actions, be directed to other objects, such as engage the attention and the pursuit of mere men of the world, there will be no real difference between him and mere

men

men of the world. Pleasure, ambition, or gain, will be equally their principal objects, those for the sake of which they would sacrifice every thing else.

Christianity does not operate as a charm. The use of it does not resemble that of a badge, or a certificate, to entitle a man to any privilege. It is of no use but so far as it enters into the sentiments, contributes to form the habits, and direct the conduct, of men; and to do this, it must really occupy the mind, and engage its closest attention; so that the maxims of it may instantly occur the moment that they are called for; and therefore in whatever it be that the true christian and the mere man of the world really differ, the difference could not fail to appear. If there was any gratification or pursuit, that did not suit the christian character, though others might indulge in it without scruple, and despise all who did not; the true christian would be unmoved by such examples, or such ridicule. His habitual fear of God, and his respect for the commands of Christ, will at all times render him superior to any such influence. Whatever his christian principles called him to do, or to suffer, he would be at all times ready to obey the call.

For any principles to have their practical influence, they must at least be familiar to the mind, and this they cannot be unless they be voluntarily cherished there, and be dwelt upon with pleasure, when other objects do not necessarily obtrude themselves. Consider, then, how many objects are perpetually occupying the minds of men in the present

state

state of things in the christian world, and how forcible their hold is upon them, and consequently how difficult it must be to prevent their all prevailing influence, to the exclsion of that of christianity.

I. The age in which we live, more than any that have preceded it, may be said to be the age of *trade and commerce*. Great wealth is chiefly to be acquired by this means. It is, at least, the most expeditious way of acquiring a fortune, with any regard to the principles of honour, and honesty. But to succeed to any great extent in mercantile business of any kind, especially now that such numbers of active and sensible men are engaged in the same, a man must give almost his whole attention to it, so that there will be little room for any thing else to occupy his mind. If he do not literally, in the language of scripture, *rise up early, and sit up late*, it will occupy his thoughts when his head is upon his pillow. His anxiety will often keep him awake. Even at that season of rest he will be considering whether it will be prudent to make this or that purchase, whether this or that man may be safely trusted, whether there will not be too much hazard in this or that undertaking, and a thousand things of this nature.

If such a person's business allow him any leisure, he is fatigued, and wants amusement, and cannot bear any thing that makes him serious. He therefore, engages in parties of pleasure, and various entertaiments, that even more than business exclude all thoughts of religion. And in this course

of

of alternate business and mere amusement or feasting, do many men of business proceed day after day, and year after year, till christianity is as foreign to their thought as if they had been heathens.

If the man of business have any turn for reading, and that not for mere amusement, it is history, or politics, something relating to the topics of the day, but not the Bible that he reads. To this, if he have not read it at school, many a man of business is an utter stranger; and though in this book God himself speaks to men, concerning their most important interests, their duties here, and their expectations hereafter, they will not listen even to their maker. On Sundays, which the laws of most christian countries prevent men from giving to business, many never go to any place of christian worship; but to relieve themselves from the fatigues of the week, make that their day of regular excursion, in company with persons of similar occupations; and their conversation, if not irreligious and profane, is at least on topics altogether foreign to religion.

II. The business of *agriculture* is much less unfavourable to religion and devotion, It does not occupy the mind in the same degree; and it is attended with much less anxiety. Nay the principal causes of anxiety to the cultivator of the ground, viz. the uncertainty of the seasons, and the weather, rather lead the thoughts to God, the author of nature, and of all its laws; from which he expects every thing that is favourable to his employment; and he passes his time in the constant view of the works of God; so that they must in some measure engage his attention. And if he attend at all to the objects with which he is continually surrounded, they must excite his admiration and devotion. This at least, is their natural tendency; though even here other objects, and other views, foreign to his proper employment, may interfere, so that, in the language of scripture, *seeing he shall not see, and hearing he shall not understand*; and giving more attention to gain than to his employment in any other view, even the farmer may be as destitute of religion as the tradesman; and great numbers, no doubt, are so. This however is by no means owing to their employment, but to other influences, which affect all men alike, without distinction of classes or ranks. This employment I therefore consider, as of all others, the most favourable to the temper and spirit of christianity.

III. In this advanced state of the world, and of society, the profession of *law* and *medicine* require more study and time than formerly. Laws are necessarily multiplied, and cases more complicated. The study of medicine requires more knowledge of various branches of science, as natural philosophy, chemistry, and botany, besides a knowledge of the learned languages, and other articles with which no physician of eminence can be unacquainted. Whether it be owing to these circumstances, or to any other, it is remarked in England, and I believe in Europe in general, that but few either of lawyers or physicians are men of religion,

religion, though some few are eminently so. Physicians have an obvious excuse for not regularly attending places of public worship; and if men can spend the Sundays without any exercise of religion, the whole week will generally pass without any, and the subject itself will find little place in their thoughts.

IV. The times in which we live may, in a very remarkable degree, be said to be the age of *Politics*, and from the very extraordinary state of the world it is in some degree necessarily so. Greater events are now depending than any that the history of any former age can shew; and the theory and practice of the internal government of countries, the circumstances that tend to make governments stable, and the people prosperous and happy, concerning which there is endless room for difference of opinion, occupy the thoughts of all men who are capable of any reflection. No person can even read the common newspaper, or see any mixed company, without entering into them. He will, of course, form his own opinion of public men and public measures; and if they be different from those of his neighbours, the subjects will be discussed, and sometimes without that temper which the discussion of all subjects of importance requires. Consequently, the subject of Politics, in the present state of things, is with many as much an enemy to religion, as trade and commerce, or any other pursuit by which men gain a livelihood. Many persons who read find nothing that interests

them

them but what relates to the events of the time, or the politics of the day.

This state of things might lead men to look to the hand of God, and a particular Providence, which is evidently bringing about a state of things far exceeding in magnitude and importance, any thing that the present or any former generation of men has seen. And a person of an habitually pious disposition, who regards the hand of God in every thing, will not take up a newspaper without reflecting that he is going to see what God has wrought; and considering what it is that he is apparently about to work. To him whatever wishes he may, from his imperfect view of things, indulge himself in (which however will always be with moderation and submission) all news is good news. Every event thas has actually taken place, as it could not have been without the permission (which is in fact the appointment) of God, he is persuaded is that which was most fit and proper for the circumstances, and will lead to the best end; and that though for the present it may be calamitous, the final issue, he cannot doubt, will be happy.

But mere men of the world look no farther than to men, though they are no more than instruments in the hand of God; and consequently, as the events are pleasing or displeasing to them, promising or unpromising, their hopes and fears, their affections or dislikes, are excited to the greatest degree; so as often to banish all tranquility of mind,

and

and cool reflection. And certainly, a mind in this state is not the proper seat of religion and devotion. All the thoughts of such persons are engaged, and their whole minds are occupied by objects, which not only exclude christianity, but such as inspire a temper the very reverse of that of a christian, which is peculiarly meek, benevolent, even to enemies, and heavenly minded, a disposition of mind which we should in vain look for in the eager politician of these times.

As to those who are concerned in conducting the business of politics, those in whose hands God has more immediately placed the fate of nations, it is not to be expected (though there are noble exceptions) that they will be eminent for piety and religion, or have any other objects than those of ambition, and, often that of avarice. Their eagerness to get into power, their jealousy of all their opponents who wish to supplant them at home, and their negociotions with foreign powers, which must be intricate, must often keep their minds upon the rack, to the exclusion of every sentiment, not only of religion, but even of common justice and humanity. For such all history shews to have been the character of the generality of statesmen and warriors, in all ages, and all nations. They have kept the world in the same state of ferment and disorder with their own minds. The consolation of a christian, in this state of things, is that the great Being, whose providence statesmen seldom respect,

respect, does, though with a hand unseen, direct all the affairs of men. *He ruleth in the kingdom of men, and giveth them to whomsoever he pleases;* and even the Pharaohs, and Nebuchadnezzars, are as useful instruments in his hands as the Davids, and the Solomons.

V. It might be thought that *philosophers*, persons daily conversant in the study of nature, must be devout; And the poet Young says *an undevout astronomer is mad*; Yet we see in fact that men may be so busy all their lives in the investigation of second causes, as intirely to overlook the great first cause of all, and even to deny that any such Being exists. Or seeing no change in the course of nature at present, or in any late period, they hastily conclude that all things have ever been as they now are from the beginning; so that if the race of men had a maker, he has ceased to give any attention to them, or their conduct; and consequently that they are at full liberty to consult their own interest, and live as they please, without any regard to him. Also philosophers, having all the passions of other men, the same love of pleasure, the same ardour of ambition, and the same attachment to gain, that actuate other men, they have in these respects been, in the usual course of their lives, governed by passion more than reason, and have lived as much *without God in the world*, as thoughtless of his being, perfections, and providence, as other men.

VI. Even ministers of the christian religion, though necessarily employed in the public offices

of

of it, and in teaching the principles of it to others, are not necessarily influenced by them themselves; though the character they sustain in society obliges them to greater external decency of conduct; so as to lay them under some considerable restraint, at least will respect to a love of pleasure, and a taste for amusement. But if the profession was not the real object of their choice, from a sense of its superior excellence, even this duty may be discharged as any other task, as any other means of subsistence, or on account of some other advantages to be derived from it. In some cases, in which religion is supported by the state, and ample emoluments are within the reach of churchmen, the christian ministry (if in such a case it can be so called) may be chosen as the means of gratifying men's ambition or avarice.

In this state of thing can we wonder at the progress of infidelity? Those who are entire strangers to it see that it has little influence on the hearts and lives of those with whom they converse, so that whether it be true or false, they think it to be of little consequence, and not worth the trouble of a serious investigation. And many persons who had nominally christian parents, giving no more serious attention to christianity than they see their parents and others give to it, observing none of its exercises, or only in the most superficial manner, seldom attending public worship, never reading the scriptures, or any book relating to religion, either explaining its evidences, or enforcing its duties, which

they

they find to interfere with their inclinations, get a dislike to the subject; and in this state of mind a mere cavil, or a jest, such as are to be found in the writings of Voltaire, and other modern unbelievers, has the force of argument. With many persons too in the upper ranks of life, christianity being the belief of the common people, on whom they look down with contempt, has more weight in their rejection of it than they will acknowledge, or than they may even be aware of themselves.

Now, as I observed before, christianity, though not absolutely and expressly rejected, is of no *use* unless it influence the temper of our minds, and our conduct in life; if it lays no restraint on the love of pleasure, the love of gain, or the pursuits of ambition, but leaves men as worldly minded in all respects as those who never heard of it; as much as if they had never heard of that future state which is brought to light by it, and which in the gospel is held up as a constant and most interesting object of attention and contemplation to all christians. We should never forget that religion is only a *means* to a certain *end;* and if we do not make this use of it, it would have been better for us never to have had it, or to have known it; since then we should have had one talent less than we now have to be accountable for. And if it be true that God has revealed his will to men, and sent messenger after messenger to promote the virtue and happiness of his rational offspring, he knew that such an extraordinary dispensation was necessary for us, and we cannot be

innocent

innocent if we neglect to attend to it, and to make the proper use of it; unless we be so situated, as never to have heard of it.

Such are the general causes of the prevailing inattention to the subject of religion; and which extinguishes in so great a degree the genuine spirit of christianity. These, therefore, in proportion to the value we set upon our religion, and in proportion to the concern we have for our own improvement and that of others, we must endeavour, by every means in our power, to counteract; *exhorting one another daily while it is called to day*, lest we be carried away by the baneful torrent, which we see to be in danger of deluging, as we may say, a great part of the nominally christian world.

The means by which this may be done are sufficiently obvious. It is the application of them only that, in such an age as this, has any real difficulty in it. And certainly it requires no small degree of fortitude and resolution to appear so singular as a sincere and zealous christian must some times do among persons of a different character. He must be content to be thought *righteous over much;* to be considered as a man of a weak mind, and devoid of spirit, and of those qualities which recommend men to the admiration of the world. For though virtue, as it is commonly understood, has the sanction of general estimation, and persons accounted vicious are universally censured; the virtues that are most admired are not always christian virtues, and give more indulgence to the passions, as to those of revenge, and a love of what is called pleasure, of various kinds, than christianity allows. And there is not perhaps any vice besides that of a mean selfishness, that is equally condemned by christianity and the voice of the world. We see that even murder, in the form of a duel, passes without any censure at all. Nay, the spirit with which men fight duels is applauded; while the meekness, though it be real magnanimity, showing a due command of temper, which overlooks insults, and preserves a kindness for those who offer them, is branded as meanness of spirit. Voluptuousness to a really criminal excess passes with so light a censure, that when any person is said to be *no man's enemy but his own*, he is not thought at all the worse of on that account, especially as it is often accompanied with a contempt of money, and a love of society like his own. Profaneness is too generally considered as no vice at all, but only at the worst a foolish and unmeaning custom.

In these circumstances, a profound reverence for the name and attributes of God, the great duty of *not living to ourselves*, but of the appropriation of the whole of a man's time, fortune, and ability of every kind, to the good of others, the love of God with the whole heart, and our neighbour as ourselves, including in the word neighbour every person to whom it is in our power to render any service; the obligation of sacrificing every thing in life, and even of life itself, for the sake of conscience,

ence, in the cause of truth and right, with a view to a recompence not in this world but another, which christianity requires of us, are things quite above the comprehension of mankind in general. And whatever men cannot attain themselves, they think to be romantic and absurd, a kind of quixotism in morals, and a just object of ridicule and contempt.

Since, then, what is called *the world*, and the prevailing maxims and customs of the times in which we live, give us no assistance, but must operate as an impediment in our christian course, we must surmount this great difficulty by our own voluntary exertions, taking to our aid those helps by which christian principles are most effectually impressed, and kept in view. Something of this kind is absolutely necessary, because no end can be gained without employing the proper means; and if any thing that does not necessarily obtrude itself requires to be *attended* to, it must be purposely brought before the mind by reflection, reading, or conversation; to do this most effectually, some *time* must be set apart for the purpose. Also those intervals of time which are not engaged by necessary business should not be wholly given to mere amusement (though something of this kind is necessary for such beings as we are) but be employed to some serious purpose.

David said that he meditated upon God *in the night watches, and upon his bed*. In the law of God, he says that a good man will *meditate day and night*. And whatever it be that we really take pleasure in,

it

it will naturally occur to our thoughts when they are not necessarily occupied by other things; being the most pleasing subjects of contemplation. The first exercise therefore that I would recommend to all christians is the frequent reading of the scriptures.

Christians have far more, and more interesting, subjects of contemplation and meditation than David had. We see much farther than he could into the great plan of providence, respecting the present and future condition of man. We are acquainted with many more instances of his intercourse with mankind, with more communications of his will, and a far more clear and explicit account of his designs respecting them. And what can be more interesting to man than his intercourse with his maker, the great Being on whom we constantly depend, *for life, breath, and all things*, who is also our moral governor, and our final judge?

Since the time of David there has been a long succession of prophets, and especially the appearance of the greatest of all the prophets, Jesus Christ; who *brought life and immortality to light*, having not only given us certain information concerning a resurrection, and a future state, but exemplifying his doctrine in his own person, by actually dying and rising from the dead. There was also a most glorious display of divine interpositions in the time of the apostles, by which our faith in the gospel is abundantly confirmed, and our attention to a future state so much excited, that it might al-

most

most have been feared, that mankind would think of little else, and that the business of this life would have been too much neglected. For what is the interest we take in all other histories compared to our interest in this? Other histories are no doubt, instructive; but the books of scripture, besides being infinitely more curious, and interesting, (as the transactions of *God*, compared with those of *men*,) may be said to be a title to an estate, to which any man may become an heir. In the scripture we are informed of the certainty, and the value, of this great inheritance, and with the terms on which we may secure the possession of it. The books of scripture are also the most ancient writings in the world, and penned with a simplicity of which we have no other example so strikingly beautiful; and they exhibit the manners of the primitive ages of mankind; so that there is in them every thing that can interest curiosity, as well as impart the most important information.

If, however, notwithstanding these recommendations, the scriptures, and other works illustrative of their contents, have not engaged the attention, it behoves every person who really wishes to imbibe the spirit of christianity, to make himself well acquainted with them, and to persist in the reading and study of them, till he find himself interested in their contents, and imbibe the pious and benevolent temper which is so conspicuous in the writers. And how irksome soever, through disuse, and other causes, the reading of the scriptures, and of other books which

which have the same tendency, may for some time be, perseverance will overcome it; and then, if I may speak from experience, no reading will be so interesting or pleasing; and the satisfaction will increase with every fresh perusal.

This circumstance enables us to account for the peculiar pleasure that David, and other pious Jews, appear to have derived from reading the scriptures. They had few other books; so that if they read at all, they must have read them perpetually in their own houses, as well as have heard them constantly read in the synagogues, from the time that they had such places of public worship, which they certainly had from the time of the Babylonish captivity. At this day there are so many other books to engage the attention, that in too many cases they totally exclude the reading of that which is of infinitely more value than all the rest.

But whatever be the leisure that any person can command for reading, some portion of it should by all means be appropriated to that kind of reading the object of which is to increase the knowledge which relates to our profession as christians. And this will lead to a course of reading both curious and interesting, especially such as makes us acquainted with the progress of christianity in the world. No kind of reading tends so much to counteract the influence of the world, and its principles, as the lives of eminent christians; and most of all the martyrs, whose piety, patience, and fortitude, in chearfully abandoning life, and every thing in it, for the sake

of

of conscience, cannot fail to inspire something of the same excellent spirit; and this once fully imbibed, will enable a man to behave as becomes a christian in every situation, of prosperity as well as of adversity, in life or in death.

Compared to the strong feelings with which such works as these are read by persons who have acquired a true relish for them, all other reading is perfectly insipid, and a truly pious christian, who has few books besides the Bible, has little cause to envy the man of letters, in whose ample library the Bible is not to be found. What is there of pathetic address in all the writings of the admired antients compared to the book of Deuteronomy by Moses? And what is all their poetry compared to the Psalms of David, and some parts of Isaiah? And yet such is the power of association and habit, that by persons of a different education, and turn of mind, those parts of scripture which are by some read with emotions of the most exalted and pleasurable kind, will be perused with perfect indifference, and even disgust: and if such persons be advanced in life, so that their habits are confirmed, the endeavour to communicate to them a relish for such writings will be altogether in vain. Of such persons we may say with Bacon's brazen statue, *Time is past.*

So strongly is my mind impressed with a sense of the importance of the habitual reading of the scriptures, both from considering the nature of the thing, and from the best attention that I have been able

able to give to particular characters and facts, that I do not see how those persons who neglect it, and who have no satisfaction in habitually meditating on the infinitely important subjects to which they relate, can be said to have any thing of christianity besides the name. They cannot feel the influence of its doctrines, its precepts, or its motives, when they give no attention to them; and therefore they cannot derive any advantage from christianity, except such as accrues to all the nominally christianized part of the world, in improving the general character, manners, and customs of it; but which, as it has not arisen from any attention that they have given to it, cannot entitle them to the character, or rewards of true christians; such as those who have lived as pilgrims and strangers here below, and as citizens of heaven; who, though living in the world, have had their affections on things above, whose treasure, the object of their chief care and pursuit, has been not in the things of this world, but in heaven. They may not be rejected by Christ as *workers of iniquity*, but they have no title to the appellation of *good and faithful servants*, to a master whom they have never truly loved or respected, and hardly even thought of; and therefore cannot expect to partake in the joy of their Lord.

II. Besides other obvious uses of public worship, a person who wishes to cultivate the true spirit, and acquire the proper habits of religion, must not neglect it. We are social beings, and our joining in any scheme in which we are alike interested,

is

is a mutual encouragement to persevere in it, and to pursue it with proper ardour. It likewise operates as a tie not lightly to desert the profession, and such a tie men concerned in the multifarious business of this life often want.

III. Private and habitual devotion is the life and soul of all practical religion, No man can be truly religious who does not, in his daily thoughts, respect the presence and government of God, and who does not regard him as the author of all things, and the sovereign disposer of all events; so as to *live as seeing him who is invisible;* as I have explained pretty much at large in a printed discourse on this subject.

IV. Family prayer, if not of absolute necessity, is of great use in all christian families. Dr. Hartley, one of the most judicious, as well of the most pious of men, says *Observations on man vol.* 2. p. 336, "I believe it may be laid down as a certain "fact, that no master or mistress of a family can "have a true concern for religion, or be a child of "God, who does not take care to worship God by "family prayer. Let the observation of the fact "determine." I would not chuse to express myself quite in this manner, since much must be allowed for the different circumstances of families; but thus much may certainly be ssid with truth, that if the practice of family prayer, or any other mode in which we give evidence to the world that we are christians, be forborn through shame, or a compliance with the modes of the world, we have no just claim to the title and privilege of christians, but fall under the awful sentence of Christ, *If any man be ashamed of me, and my words, in this generation, of him will the son of man be ashamed when he comes in the glory of his father, with the holy angels*, Mark viii. 38.

Every practice by which we declare our belief of christianity, such as attending christian worship, receiving the Lord's supper; or performing any other acknowledged christian duty, tends to strengthen our faith, to inspire the proper spirit of the profession, and secure the performance of every duty which it enjoins; and therefore should by no means be neglected by us.

Thus should we be urgent, even to *exhort one another*, and all should gladly and thankfully *receive the word of exhortation;* to *be steadfast, immoveable, always abounding in the work of the Lord, knowing that our labour will not be in vain in the Lord.*

The author of this epistle says (Ch. x. 25.) We should exhort one another *so much the more as we see the day*, meaning, no doubt, the great day or the second coming of Christ, *approaching*. If this motive had weight in the times of the apostles, it must have more now; since that great day, which *will try every man's work what it is*, must be nearer than it was then; and though this time was not known to our Lord himself, but only the signs of its approach, many intelligent christians, who are

attentive

attentive to the *signs of the times*, are of opinion that it cannot now ce far distant, and may be expected even in the present generation. But since the coming is certain, though the time be uncertain, let us be *ready*, that *when our Lord, shall return, and take account of his servants, we may be found without spot, and not be ashamed before him at his coming.*

ON

ON

FAITH AND PATIENCE.

These all died in faith, not having received the promises, but having seen them afar off; and were persuaded of them, and embraced them, and confessed that they were strangers and pilgrims on the earth.

Heb. xi. 13.

THE great use of religion is to enlarge the minds of men; leading them to look beyond themselves, and beyond the present moment; to take an interest in the concerns of others, and to look forward to the most distant times. By this means men become less selfish, and at the same time more intellectual; being less governed by the impulses of mere sensual appetite, which is the characteristic of brutal nature, and also of a state of childhood.

This habit of mind cannot be imparted by instruction. It must necessarily be the fruit of experience. And since this advance in intellectual improvement implies the forbearance of immediate gratification, which is always painful, a state of suffering is an essential ingredient in this important discipline of the mind, and therefore ought not by

any

any means to be complained of, by those who wish not to retard their progress towards perfection.

We see in the affections and conduct of children how injurious constant indulgence is to them, and how necessary to their own future happiness, as well as to the comfort of those who are about them, are frequent checks and restraints. The less is the gratification of their wishes restrained, the more eager are their desires, and the more confident their expectation of any desired event; and consequently the more painful is disappointment to them. And since disappointment will necessarily come, from the absolute impossibility of gratifying all their absurd wishes, the more they must suffer from impatience and vexation in consequence of a want of early checks.

It is happy for men that, in a state of infancy, they cannot explain their wants; so that whatever they feel or wish, it has little or no connection with what they experience. They must necessarily be many years under the absolute government of others. This lays a foundation for a habit of patience and forbearance, which is of infinite value to them, and which must be carried much farther as they advance in life, if they advance in intellectual and moral improvement.

We see not only in the case of indulged children, but in that of kings, and others who have many persons intirely subservient to them, that a habit of indulgence makes them incapable of brooking disappoint-appointment; so that they suffer infinitely more than persons who frequently meet with them, and who have by that means acquired a meek disposition, and a habit of patience and forbearance. These persons can enjoy the pleasures of life without suffering much from the evils of it; whereas they who have not been in a situation proper for acquiring this habit, not only suffer much from evil; but have little enjoyment even of good. This being nothing more than they always expect, and what from frequent indulgence they receive with much indifference, often bordering on disgust.

Hence it follows that, in exercising the faith and patience of men, God acts the part of a kind and judicious parent, attentive to the improvement of his children; not affected by their present temporary feelings, but consulting their happiness at a future period, and in the whole of their existence; this life, long as it may be, being only the infancy of man, in which are to be formed habits that are to qualify them for superior and more lasting enjoyment hereafter. Compared to eternity, what is time? what is the longest term of human life? If the whole of it should be passed in suffering, there is room for an abundant recompence in a future state. But our merciful father has given sufficient proof of his benevolence in the provision that he has made for the enjoyment of this life, happiness greatly exceeding the misery that is so much complained of in it. From this his disposition, and his wish,

wish, to make his offspring happy is sufficiently evident; and we have just ground to hope, and believe, that all the sufferings of this life are in their nature preparatory to our happiness in another, provided they have their proper effect upon our tempers and dispositions.

We see most of the conduct of divine providence in the scriptures, which are eminently calculated for our instruction; and we there see that the methods of the extraordinary providence of God, in his intercourse with mankind, is exactly correspondent to the plan of his general providence. We there see that from the beginning of the world he has been training men to virtue and happiness by a course of severe but salutary discipline; some of the most eminent of our race, those whom we may call the greatest favourites of heaven, with whose history we are best acquainted, having been treated in such a manner as to exercise their patience to the utmost, before they were distinguished by any reward for it. As an attention to particular cases, such as are briefly recited in the eleventh chapter of the epistle to the Hebrews, will be eminently instructive, I shall enlarge a little on some of them, noticing such circumstances in their history as appear to be most remarkable.

Abraham, at the age of seventy-five, was command by God to leave his native country, on a promise that he would give him another which he would shew him, and that he would make his descendants a great nation. Accordingly, he left Chaldea,

dea, and went to Haran in Mesopotamia and the year following he proceeded to the land of Canaan, Gen. xii. 4. There God appeared to him the second time, telling him that that was the country destined for him. Ten years, however, passed without the appearance of any issue, from which the promised nation was to descend; and in the mean time he had been obliged by a grievous famine to go into Egypt.

At his return the promise of his descendants becoming a great nation was renewed, and again, in a peculiarly solemn manner, after his rescue of Lot; but having no hope of any son by his wife Sarah, he was prevailed upon by her to take her maid Hagar, and by her he had Ishmael, when he was eighty-six years old. But this was not the son from whom the great nation was to descend; and it was not till he had arrived at the advanced age of ninety-nine that he was promised to have a son by Sarah, who was then ninety; so that her conception was out of the course of nature. Notwithstanding this long delay and the most unpromising appearances, his faith did not fail; and on this account he was highly approved by God, Gen. xv. 6. Accordingly he had a son the year following, but only one; so that, to all appearance, his having a numerous posterity was very uncertain.

To give the greater exercise to his faith, when this son, so long expected, was arrived at years of maturity, the affectionate father received a command from God to sacrifice him; a command which

which he hesitated not to obey, though to appearance this act of obedience would put an end to all his flattering prospects. This, however, was merely a trial of his faith, and the order to sacrifice his son was countermanded.

When Isaac was forty years old, and his father one hundred and forty, he was married; but twenty years more elapsed before he had a son, so that Abraham was one hundred and sixty years old, and saw no more than two grand children, and when they were boys of fifteen he died. His expectation, therefore, of a numerous posterity could not have arisen from any thing that he saw, but altogether from his faith in the divine promise.

After this the hopes of the family, were limited to Jacob one of the sons of Isaac; and he did not marry till he was near fourscore years of age, and at his outset he appeared to have been greatly inferior to his brother. For when he returned from Padan Aram no mention is made but of his wives, his children, and his cattle, whereas his brother met him with four hundred men, and made very light of the very valuable present that Jacob forced upon his acceptance.

In the family of Jacob we see, however, at length, the rudiments of a clan, or nation; and when they went into Egypt they mustered seventy males, but their situation in servitude, to which they were soon reduced, was very unpromising with respect to any future greatness. The life of

Jacob himself had little in it to be envied. After leaving his parents, where though he was the favourite of the mother, he was by no means so of the father, he served his uncle Laban twenty years; and by his own account he underwent great hardships, and was grievously imposed upon. At his return he suffered much from the fear of his brother's resentment. The behaviour of several of his sons must have been a source of much affliction to him, and the loss of Joseph must have gone near to break his heart. In this state he continued fifteen years, when near the close of his life he was comforted by the recovery of his favorite son, and the settlement of all his family in a plentiful country. But though he knew, from the warning that God gave to Abraham, that his descendants would soon be reduced to a state of great oppression, and would continue in it many years, he died in the firmest faith that they would in future time become a great and flourishing nation; and he distinctly foretold the fate of each of his sons, as the heads of great tribes, of which that of Judah would be the most distinguished.

Joseph, the most pious and virtuous of his sons, was exercised in the severest manner. After being the favourite of his father till he had attained the age of seventeen, he was sold for a slave; and, in consequence of a false accusation, confined in prison several years. But these unfavourable circumstances were probably those that contributed most to the

the peculiar excellencies of his character; disposing him to be humble and serious, wholly resigned to the will of God; and believing that his providence had the disposal of every thing, he entertained no sentiment of revenge on account of the injuries that had been done to him. Looking forward to the future greatness of his descendants, and confiding in the divine promise, that the family would become possessed of the land of Canaan, he ordered that he should not be buried in Egypt, but be embalmed, in order to be carried to the promised land when they should remove thither.

Though the descendants of Jacob multiplied greatly in Egypt; yet no person, seeing the state of abject servitude to which they were there reduced, could have imagined that they were destined to rise superior to their proud masters, and make the figure they afterwards did under David and Solomon, and much less that they would become the most distinguished of all nations, which if the predictions concerning them have their accomplishment, they are to be. The Israelites in general seemed to have abandoned all hopes of the kind, and to have acquiesced, through despair, in their servile condition.

Moses, their future deliverer, fled from the country at the age of forty, and continued forty years more among the Arabs, where he married, and evidently never thought of returning to join his brethren; when the divine Being appeared in a most extraor-

extraordinary manner in their favour, delivering them as it is said, *with a high hand and an outstretched arm*, from the power of the Egyptians, at a time when there were no visible means of accomplishing it.

But though the nation was in this extraordinary manner delivered from their state of servitude in Egypt, yet, wandering as they did no less than forty years in the wilderness, surrounded by warlike nations, they could not, except in reliance on the divine favour by which they were conducted, have expected to make the conquest of such a country as Palestine then was, fully peopled, and by nations in the habits of war, with all their considerable towns fortified. Yet in this manner was the favourite nation training up for their future greatness, when, to an indifferent spectator, their condition would have appeared very uncertain and hazardous; not likely to make any greater figure than one of the hordes of Arabs, and having nothing but the very worst and least cultivable part of Arabia to settle in; every fertitle spot in the country being already occupied.

The people in general at this time thought so ill of their situation and prospects, that nothing but very extraordinary interpositions in their favour oould have prevented their returning into Egypt, which they again and again wished to do. The faith, however, of the more pious among them never failed; and after the expiration of the forty years they were put into the possession of a considerable

tract

tract of country on the East of the river Jordan. But at this time not only were the descendants of Esau a well settled and considerable nation, but even those of Moab and Ammon, the two sons of Lot, though they were destined to bow to the superiority of the wandering Israelites.

After they got possession of the land of Canaan, in a manner as extraordinary as their emancipation from their bondage in Egypt, they made no considerable figure for the space of about four hundred years; and during a great part of it they were in subjection to some or other of the neighbouring nations, in consequence of their apostacy from their religion; so that in all this time there was far from being any appearance of their being what they were in the reigns of David and Solomon; and this state of prosperity did not continue quite a century. After this they relapsed into their former inconsiderable state, and they were finally conquered, and carried into captivity, by the kings of Assyria and Babylon; when to all appearance there was an end of the nation of the Israelites, as there was to those of the Moabites, Ammonites, and Philistines, which never rose to any degree of power or independence.

Of all the kings of Israel, David, whose piety was most exemplary, though, from the strength of his passions, his failings were very great, was exercised with the greatest trials, both before he was king and afterwards, of which many of his psalms, composed in a mournful strain, are a sufficient evidence.

dence. He was anointed king of Israel when he was very young; but though he soon distinguished himself as a warrior, he was immediately exposed to the jealousy and persecution of Saul; so that during the remainder of his reign he was obliged to take refuge in the neighbouring countries; and after the death of Saul he was seven years at Hebron, acknowledged by the tribe of Judah only.

On the other hand, Solomon, who had, no doubt, every possible advantage of education, and arrived at the most splendid situation without any difficulty, was not only excessively luxurious, but swerved from his duty in an article with respect to which his firmness might have been least of all suspected; not only indulging his wives in the idolatrous worship of the countries from which he had taken them, but joining them in it.

After this seeming annihilation of the Israelites as a nation in the captivity by Nebuchadnezzar they were, according to express prophecies, restored to their own country, though they never rose to the height from which they had fallen; and in consequence of their relapsing into vice, though not into idolatry, and rejecting the great prophet Jesus Christ, the vengeance predicted long before by Moses came upon them to the uttermost. They were conquered by the Romans, and soon after intirely driven from their country to every part of the habitable world; and in this state they remain to this day, but they are not destroyed. They preserve their peculiar customs, and never lose sight of their relation

relation to their great ancestors, or the promises of God to them, that they are to be once more, and finally, settled in their own country, and to be the most respectable of all nations. Though they are treated with the greatest contempt by all other people, they are justly proud of their descent, and of their peculiar relation to God. Whatever be the vices with which they are chargeable, they are not deficient with respect to faith. Their most necessary virtue is fully exercised, and improved, by the severe discipline to which they have been subjected.

This is the more remarkable, as none of all their calculations, or conjectures, concerning the time of their deliverance and exaltation have been verified; so that they now desist from forming any opinion on the subject, but wait with patience for the accomplishment of the promises, notwithstanding the most discouraging aspect of things, and in perfect uncertainty will respect to the time.

The Messiah, who was first promised to them with any distinctness in the time of Isaiah, they fully expected, from their interpretation of the prophecies of Daniel, about the commencement of the christian æra, when they became subject to the Romans; a situation which they brooked very ill. Jesus was the predicted Messiah, but his first coming was not to be that glorious one with which they fondly flattered themselves. And with respect to his second coming christians themselves have their

faith

faith as much exercised as is that of the Jews. It was by many fully expected soon after the age of the apostles. After this disappointment, they fixed upon later dates; but, like the Jews, we have flattered and deceived ourselves again and again. Our faith, however, does not fail, especially as our Saviour has apprized us that the time of his second coming was not known even to himself, but to the father only; and that when it will come it will be as unexpected as that of a thief in the night.

If the faith of the founders of the Jewish nation, and that of the nation itself, has been so much exercised, that of Jesus Christ and his followers has been no less so. Christ himself was *made perfect through suffering*, Heb. ii. 10. his followers cannot reasonably expect to be trained to virtue and happiness in any other way. He was *despised and rejected of men, a man of sorrows and acquainted with grief.* During the whole course of his benevolent ministry, in which he continually *went about doing good*, he met with more opposition from the envy and malice of his powerful enemies, than if he had been the pest of society. Though he gave the rulers of his nation no cause of offence besides that of reproving them for their vices, they never ceased to persecute him till they had put him to a painful and ignominious death; and he faithfully apprized all his disciples, that if they would follow him, they must *take up their cross to do it;* and that they would be *hated of all men for his name's sake*, but

that

that they ought to rejoice in being so distinguished; since in consequence of being *persecuted for righteousness sake, great would be their reward in heaven.* If they *suffered with him, they would,* as *the apostle says, reign with him, and be glorified together.*

The apostles, and the primitive christians in general, found this to be a faithful and true warning. In following the steps of their master they were persecuted as he had been; and christians received no countenance from the powers of the world for the space of three hundred years. And after this the professors of a purer christianity (for it was never more than a corrupt species of it that was patronized by princes and states) continued to be exposed to cruel persecution in various forms. Indeed they suffered much more from nominal christian powers than they had ever done from the heathen ones. It has, therefore, been true in all times, that *through much tribulation men have entered into the kingdom of God;* and consequently whenever the world smiles upon us, there is just ground for suspicion that all is not right with us.

Looking through the history of christianity from the beginning, we shall find that the most distinguished characters, those we look up to with the greatest reverence, as patterns of piety, benevolence, and constancy, have been those who have suffered the most. This was eminently the case of the apostles in general, and especially of Paul, the most active of all the propagators of christianity. For ardour of mind, and indefatigable exertion in the cause of truth and virtue, he stands unequalled in christian history. But what did he not suffer after he embraced christianity, from the malice of the Jews, and false brethren among christians.

Speaking of some who undervalued him in the church of Corinth, he gives the following brief enumeration of his labours and sufferings, Cor. xi. 23. *Are they ministers of Christ, I speak as a fool, I am more. In labours more abundant, in stripes above measure, in prisons more frequent, in deaths often. Of the Jews five times received I forty stripes save one. Thrice was I beaten with rods. Once was I stoned. Thrice I suffered shipwreck.* And this was written before the shipwreck of which a particular account is given in his history. *A night and a day I have been in the deep. In journeyings often, in perils of waters, in perils of robbers, in perils by my own countrymen, in perils by the heathen, in perils in the city, in perils in the wilderness, in perils in the sea. In fastings often, in cold and nakedness, besides those things that are without, that which cometh upon me daily, the care of all the churches. Who is weak, and I am not weak? Who is offended, and I burn not? If I must needs glory, I will glory in the things that concern my infirmities.* After this he was imprisoned two years in Judea, conveyed to Rome as a prisoner, and suffered shipwreck at Melita. He was two years more a prisoner in Rome, though

though not in strait confinement; and though he was at that time acquitted, he afterwards suffered martyrdom.

In the same epistle, however, in which he gives this account of his sufferings, he says, 2 Cor. vii. 4. *I am exceedingly joyful in all our tribulation;* and he frequently exhorts the christians to whom he writes to *rejoice in the Lord always.* Rom. xii. 12. *rejoicing in hope, patient in tribulation.* When he was preaching to some of the churches in Asia Minor, (Acts xiv. 22.) exhorting the disciples to continue in the faith, he reminds them that *through much tribulation they must enter into the kingdom of God.*

With what true heroism and satisfaction does he reflect upon his labours and sufferings in the epistles which he wrote from Rome, towards the close of his life, and when he was in expectation of a violent death. In these circumstances he thus writes to Timothy. 2 Tim. iv. 5. *Watch thou in all things. Endure affliction. Do the work of an Evangelist. Make full proof of thy ministry. For I am now ready to be offered, and the time of my departure is at hand. I have fought the good fight, I have finished my course, I have kept the faith. Henceforth there is laid up for me a crown of righteousness, which the Lord, the righteous judge, shall give me at that day; and not to me only, but to all them that love his appearing.*

Can

Can any thing now be wanting to reconcile us to any hardships to which we can ever be exposed, either in the ordinary course of providence, or in the cause of truth and a good conscience? What is all that we can suffer, in these times of rest from open persecution, compared to that to which either the antient martyrs in the time of heathens, or those in the time of popery, were continually exposed. How many thousands of them suffered death in every frightful form, besides being the objects of ridicule and insult, as if instead of being the benefactors of mankind, they had been the greatest pests of society; a treatment which to many persons is more painful than death itself, and very often would be intolerable, were it not that the attachment of friends is a balance to the contempt of enemies,

It is true, however, that something of this kind of persecution still remains to those who resolutely bear their testimony, at the same time in favour of christianity, and against the manifold corruptions of it with nominal christians, even those who call themselves *reformed*. In this case we cannot expect to escape the ridicule of the philosophical part of the world on the one hand, and the hatred of bigots on the other. In some situations it requires no small degree of fortitude to bear this with a temper becoming christians, pitying the ignorance and prejudices of men, without bearing them any ill will; and taking every method of removing their ignorance and prejudices, in a manner the least offensive to them;

them; always joining the wisdom of the serpent, and the innocence of the dove; the seriousness of the christian, with the ease and cheerfulness of the benevolent man; free from that offensive austerity which gives many persons an aversion to religion, as if it was an enemy to human happiness, and the parent of gloom and melancholy.

Let us more particularly apply this doctrine to the great object of christian hope, the second coming of Christ with power and great glory, to raise the dead and to judge the world, when he will render to every man according to his works. We are apprized by the apostle Peter. (2 Peter iii. 3.) that in *the last days, there will be scoffers*, as we now find, *who will say Where is the promise of his coming? For, since the fathers fell asleep, all things continue as they were from the beginning of the creation.* But, as he observes, *one day is with the Lord as a thousand years, and a thousand years as one day*, that *he is not slack concerning his promise*, and that *that day will come though as a thief in the night.*

Let us then be ever *looking for*, as we are *hasting unto*, the *coming of this great day of God; and be diligent, that we may be found of him without spot and blameless.* That greatest of all events is not the less certain for being delayed beyond our expectation. The Israelites, no doubt, expected to enter the promised land immediately after their leaving Egypt; but though they passed forty years in the wilderness,

wilderness, they nevertheless were put in the full possession of it when that time of their probation was expired; so that we read, Josh. xxi. 43. *The Lord gave unto Israel all the land which he sware to give unto their fathers. There failed not one of the good things which the Lord had spoken to the house of Israel: all came to pass.* In like manner, no doubt, we shall all have occasion to say the same in due time, when our eyes, and every eye, shall see Christ coming in the clouds of heaven, be the distance of that time from the present ever so great. Let us, therefore, live as if it was near at hand. With this prospect before us, *what manner of persons,* as the apostle Peter says, *ought we to be in all holy conversation and godliness.*

But, as individuals, we have no occasion to enter into any speculations about the time of this greatest of all events, in which we are so much interested. To each of us it must be very near. For since we have no perception of time during a profound sleep, we shall have none while we are in the grave. The sleep of Adam will appear to him to have been as short as that of those who shall die the day before the second coming of Christ. In both cases, alike, it will be as a moment; so that our resurrection will seem immediately to succeed the closing of our eyes in this world. What a sublime and interesting consideration is this. For *what is our life,* but, as the apostle says, *like a vapour, which appears for a little time and then vanishes away;* and immedi-

immediately after this the great scene opens upon us. May we all be so prepared for it, that when our Lord shall return, and *take account of his servants, we may have confidence, and not be ashamed before him at his coming.*

ON

ON

THE CHANGE WHICH TOOK PLACE

IN THE CHARACTER OF THE APOSTLES

AFTER THE RESURRECTION OF

JESUS CHRIST.

[PART I.]

And when they saw the boldness of Peter and John, and perceived that they were unlearned and ignorant men, they marvelled, and they took knowledge, of them that they had been with Jesus.

Acts. iv. 13.

THERE is nothing in all history, and certainly nothing within the compass of our own observation and experience, that shows so great a change in the views and characters of men, as we find to have taken place in the apostles after the resurrection and ascension of Jesus, or rather after the descent of the Holy Spirit on the day of Pentecost. They appear to have always been honest, virtuous and pious men; but having imbibed the prejudices of their nation, they expected a temporal prince in their Messiah; and supposing their master to be that Messiah, and being in favour with him, they, with the ambition that seems to be natural to all men,

men, hoped to be advanced to the first places in his kingdom, and, seemingly, without considering whether they were qualified to fill them or not.

With these views, and no higher, they attached themselves to Jesus, after being convinced by his miracles that he was a true prophet; and conceived the idea, though without its having being suggested by himself, that he was the Messiah they were looking for. They had frequent disputes among themselves on this subject; and two of them were so impatient, and presumed so much on their superior merit, that, without regarding the offence it would necessarily give to the other apostles, they actually applied to Jesus for the distinction of sitting the one on his right hand and the other on his left, when he should be in the possession of his kingdom.

Though Jesus never failed to repress these ambitious views, and never gave the least encouragement to them in any of the apostles, not even in Peter, whose pretensions seem to have been the best founded, they all retained this idea till the time of his death. This event so contrary to their expectations, disconcerted and confounded them, and necessarily obliged them to give up all their fond expectations of worldly preferment. But after his resurrection their ambition revived, and they could not forbear to ask him (Acts i. 6.) if he would then restore the kingdom to Israel, expecting, no doubt, to share in the honours and emoluments of it.

That he was destined to be a king, and that they were to partake of the honours of his kingdom, he had never denied. Nay he had given them positive assurance of it, saying (Matt. xix. 28.) that "when "he should sit upon the throne of his glory, they "should also sit upon twelve thrones, judging the "twelve tribes of Israel." But at the same time he gave them sufficient intimation that his kingdom was not to resemble the kingdoms of this world, in which the great mass of the people were subservient to the gratification of a few. For that, on the contrary, the persons the most distinguished in his kingdom would be those who should be the most assiduous to promote the happiness of others, or that they would be in fact in the capacity of *servants*, as he himself in reality was.

Whether they clearly understood his meaning does not appear, but it is probable they did not. For still their chief expectations were confined to the honour and advantage that would accrue to themselves, without attending to any obligation they would be under to promote the good of others. Whatever was meant by this kingdom, in the honours of which they were to partake, he never gave them any information concerning the *time* of its commencement. Nay, he expressly told them that this was not known even to himself. After his resurrection he professed the same ignorance, and, repressing their curiosity on that subject, he said "it "was not for them to know the times and seasons "which God had reserved to himself. (Acts i. 9.)

Reflecti-

Reflection, however, on the death of their master, on his resurrection and ascension, without his having given them any promise of his speedy return, and the recollection of the persecutions to which he had constantly apprized them they would be exposed, as that "they would be hated of all men for "his name's sake, and that they who should kill "them would think they did God service," could not fail to satisfy them that they had nothing of advantage to look for in this life; and therefore that the kingdom which he had promised them, and of the *certainty* of which they entertained no doubt, must be in another after death. And when, after this, they found themselves impowered to work miracles as Jesus had done, in confirmation of his doctrine, they, naturally timid as they had been before, assumed the courage of the antient prophets, no more overawed by men in power than they or their master had been, and making light of, nay glorying in, all the sufferings to which they were exposed.

This natural effect of their new situation, and new and more enlarged views, astonished their adversaries, who wondered how men in some of the lower classes of life, without fortune or education, should appear so fearless; and, without respecting any human authority, despising their threats, and their punishments, boldly preach what they thought themselves authorized by God to do, though in the most peremptory manner forbidden by them.

From this time, also, so far were they from envy-

ing

ing one another, or contending, as they had done before, about the chief places in their master's kingdom; having now no distinct idea of any difference that would be made among them hereafter, they considered one another as brethren, standing in the same relation to their common master; and being equally exposed to persecution on that account, their attachment to one another was such as the world had never seen before. Remembering at the same time the great stress that their master had laid on *brotherly love*, and the mutual kind offices that flowed from it; and considering all the things of this world as wholly insignificant in comparison with their glorious expectations in another, many of them made no difficulty, in the first ardour inspired by their situation, of giving up all their worldly property to those of their brethren who stood in need of it, in sure expectation of receiving their reward in heaven.

This most remarkable and sudden, and yet permanent, change in the temper and disposition of the apostles, and other primitive christians, furnishes no inconsiderable evidence of the truth of christianity, as it implies the fullest possible conviction in their minds of the truth of the great facts on which it depends; facts which immediately preceded this change, and must have been the proper cause of it, and they were certainly the best judges in the case. If they had not all known, to the greatest certainty, that Jesus was actually risen from the dead, and ascended into heaven, and that the pow-

ers

ers with which he had been endued were transferred to them, they must have been the same men that they were before, acting upon the same principles, and in the same manner, especially as they were not very young men, and some of them pretty far advanced in life. Consequently, their worldly ambition, and their envy and jealousy of each other, must have been the same that it had been before. Whereas now we find every thing of this kind quite changed and this change was not momentary, but continued through life with them all. The low passions and narrow views, and their consequent envy and jealousy, never returned, but they continued to the latest period of life what they appear to have been presently after the remarkable events above mentioned.

That such men as they evidently were, and especially in the middle and lower classes of life, unlearned, and so many of them, should concur in any imposture, and one so suddenly formed as their's must have been, whatever had been its *object*, cannot be supposed, and much less an object that had nothing in it that mankind in general value in this life; and especially that they should all act in such perfect harmony so long. That not one of them, though urged by the fear of death, or the hope of reward, should have made any discovery to the prejudice of their former associates, and that none of their enemies, sagacious and inveterate as many of them were, should have been able to detect

their

their imposture, adds infinitely to the improbability of its being one.

When these new and great views first opened upon the converts to christianity, when they saw their cause to be that of God, by the evidence of the miracles which supported it, and they were themselves occasionally under supernatural influence, this extraordinary fervour, and the effects of it, especially in acts of beneficence to their brethren, was natural. But as first impressions are always the warmest, this zeal would in a course of time as naturally abate, especially as miracles became less frequent, and their intercourse with the world would gradually tend to produce the same attention to the things of this world by which other persons were influenced.

In this situation many of them would require to be reminded of their great views and expectations in another world, by which they had at first been so much impressed, and to be exhorted to the virtues to which they lead. Accordingly, the apostles, seeing no doubt this unfavourable influence, and aware of the tendency and progress of it, do not fail in their epistles to warn them on the subject; and this they do with a distinctness and energy of which we find no example before their time.

And as we at this distance from the time of the first propagation of christianity, who receive all our impressions of it from reading and meditation, and especially as we live in a season of rest from all persecution (a situation which has its disadvantages as well

well as its advantages) are naturally less under the influences of its principles, and more exposed to those of the world at large, it may be useful to collect, and particularly attend to, all that the apostles have urged on this most interesting of all subjects; that we may see the firmness of their faith in the great doctrine of a resurrection and a future state, and the influence which they evidently thought it ought to have on men's sentiments and conduct.

It will also be pleasing, as well as useful, to observe the difference which these views made in the state of their own minds. What a wonderful change was produced in them after the death and resurrection of their master, so that they were no longer the same men.

I shall begin with the epistle of Peter, the chief of the apostles, but, who had, no doubt, been as much under the influence of worldly ambition as any of them, as may be suspected from his observing (Mark xvi. 28.) that "they had forsaken all" to follow Jesus, and desiring to know what they should receive as a compensation for the sacrifice; at that time, no doubt, expecting it in this life. What were his views and expectations afterwards, and to the close of a long life, we shall now see. At the same time we cannot fail to perceive a peculiar dignity and energy in the language of this apostle, worthy of the chief of them. The faith of Paul was equally strong, and led him to act with the same disinterestedness and courage, and it is probable that he went through more labour, and in the

course

course of his preaching suffered more; but his language on the same subject has not quite the same dignity, and force.

With what confidence and exultation does this apostle speak of the sure hope of christians in another world, and how justly, and forcibly, does he urge it as a motive to bear with patience and cheerfulness all the persecutions to which they were exposed, in the following passages of his epistles.

"Blessed be the God and father of our Lord Je-
" sus Christ, who, according to his abundant mer-
" cy, has begotten us again to a lively hope, by the
" resurrection of Jesus Christ from the dead, to an
" inheritance incorruptible, and undefiled, and that
" fadeth not away, reserved in heaven, for you who
" are kept by the power of God through faith unto
" salvation, ready to be revealed in the last time;
" wherein ye greatly rejoice, though now, for a
" season, ye be in heaviness through manifold
" temptations; that the trial of your faith (being
" much more precious than of gold which perish-
" es) may be found unto praise, and honour, and
" glory, at the appearance of Jesus Christ; whom
" having not seen ye love, in whom though now ye
" see him not, yet believing ye rejoice with joy un-
" speakable and full of glory, receiving the end of
" your faith, even the salvation of your souls,"
1 Pet. i. 3. &c.

"Beloved, think it not strange concerning the
" fiery trial which is to try you, as if some strange
" thing happened unto you, but rejoice, in as much

" as

" as ye are partakers of Christ's sufferings; that
" when his glory shall be revealed, ye may be glad
" with exceeding joy. If ye be reproached for
" the name of Christ happy are ye, for the spirit of
" Glory and of God resteth upon you. On their
" part he is evil spoken of, but on yonr part he is
" glorified."

" If any man suffer as a christian let him not be
" ashamed, but let him glorify God on this behalf.
" Wherefore let him that suffers according to the
" will of God commit the keeping of his soul unto
" him in well doing, as unto a faithful creator."
1 Pet. iv. 12. &c.

" Give diligence to make your calling and elec-
" tion sure. For if ye do these things ye shall ne-
" ver fail; for so an entrance shall be administered
" to you abundantly into the everlasting kingdom
" of our Lord and Saviour Jesus Christ." 2 Pet.
i. 10.

" The God of all grace, who has called us to his
" eternal glory by Christ Jesus, after ye have suffer-
" ed a while, make you perfect, stablish, strength-
" en, settle you." 1 Pet. v. 10.

" When the chief shepherd shall appear, ye shall
" receive a crown of glory that fadeth not away."
1 Pet. v. 4.

" Wherefore, gird up the loins of your mind, be
" sober, and hope to the end, for the grace that is to
" be brought to you at the revelation of Jesus
" Christ." 1 Pet. i. 13.

Well then might he say, "If ye suffer for righte-
"ousness sake happy are ye. Be not afraid of their
"terror, neither be troubled." 1 Pet. iii. 14.

With what noble magnanimity does this apostle contemplate the dissolution of the present state of things, and the commencement of the glorious one that is to follow it, adopting the language of the antient prophets in describing great revolutions in the world.

"Seeing that all these things shall be dissolved,
"what manner of persons ought we to be in all ho-
"ly conversation and godliness; looking for, and
"hasting unto, the coming of the day of God,
"wherein the heavens being on fire shall be dissol-
"ved, and the elements melt with fervent heat.
"Nevertheless, we, according to his promise, look
"for new heavens, and a new earth, wherein dwel-
"leth righteousness. Wherefore, beloved, seeing
"ye look for such things, be diligent, that ye may
"be found of him without spot and blameless." 2
Pet. iii. 11. &c.

The consideration of the *time* when the great and happy event is to take place gave him no concern, since he depended upon the *certainty* of it; and when we are dead the time of the resurrection will be a matter of perfect indifference to all of us. For whatever be the time of our death, that of the resurrection will appear to us to be contiguous to it. And the reason for the seeming delay, and of the uncertainty with respect to the time of the resurrec-

tion

tion and future judgment, are very rationally and satisfactorily given by him, on the principle of this being a state of trial and discipline, in which it behoves us to be in continual expectation and preparation for an event so infinitely momentous.

"There will come in the last days scoffers, walk-
"ing after their own lusts, and saying Where is the
"promise of his coming; for since the fathers fell
"asleep all things continue as they were from the
"beginning of the creation. But, beloved, be not
"ignorant of this one thing, that one day is with the
"Lord as a thousand years, and a thousand years
"as one day. The Lord is not slack concerning
"his promise, but is long suffering to us ward, not
"willing that any should perish, but that all should
"come to repentance." 2 Pet. iii. 3. &c.

Such is the animating and consoling language of this great apostle, addressed to his fellow christians, then in a state of persecution, which left them no prospect of peace or comfort in this life. And, surely, it must have been effectual to answer its purpose. The writings of this apostle are such as we may quote as *authority* for the truth of this great doctrine of another life, as he received it from Jesus, and it was confirmed by miracles wrought by himself, as well as by his brother apostles, who were endued with the same powers of which they had been witnesses in their common master.

The apostle John was one of the two brothers, the sons of Zebedee, whose eager ambition led them openly

openly to solicit the most distinguished honours in the kingdom of their master, though at the evident risk of giving the greatest offence to the rest of the twelve, all whose pretensions must have been nearly as good as theirs. But how changed do we find him at the time of writing his epistles. Here we are far from perceiving any marks of worldly ambition. On the contrary, no man could appear to be more weaned from any attachment to this world, or more desirous to wean others from it. "Love "not the world," says the heavenly-minded apostle, "nor the things that are in the world. If any "man love the world, the love of the father is not "in him. For all that is in the world, the lust of "the flesh, the lust of the eye, and the pride of life, "is not of the Father, but is of the world; and the "world passes away, and the lust thereof, but he "that doth the will of God abideth for ever." 1 John ii. 15. &c.

Such is the change that new views and principles can make in men. It is not now any thing in this life, which is so uncertain, but that *eternal life*, promised by Jesus, that is the object of his pursuit; and this he thus earnestly recommends to others. "This," says he, 1 John, ii. 25, "is the promise "which he has promised us, even eternal life. This "is the record that God has given us; eternal "life and this life is in his son. These things I "have written unto you that ye may believe on "the name of the son of God, and that ye may "know

"know that ye have eternal life." 1 John vi. 1. &c.

With what joyful expectation does he now look forward to the return of his master in his glory and kingdom. "Beloved, now are we the sons of God, "and it does not yet appear what we shall be, but "we know that when he shall appear we shall be "like him, for we shall see him as he is, iii. 2. And "now, little children, abide in him, that when he "shall appear we may have confidence, and not be "ashamed before him at his coming, ii. 28.

In the book of Revelation, interpreting this glory, he says Ch. i. 7. "Behold he cometh in the "clouds, and every eye shall see him, and they also "who pierced him, and all the kindreds of the earth, "shall wail because of him;" meaning, no doubt, his enemies, and by no means his friends, to whom it will be a season of the greatest joy and triumph. There, as Jesus said before, John xvi. 22, "their "sorrow will be turned into joy. Now ye are in "sorrow, but I will see you again, and your heart "shall rejoice, and your joy no man taketh from "you." Then the glory which his father gave to him he will give to them, xvii. 22.

James, the other ambitious brother, was the first of the apostles who died a martyr to christianity, being beheaded by Herod Agrippa, fourteen years after the death of Christ; so that there cannot be a doubt but that he had abandoned all views to advancement in this world, as well as the rest of the apostles. Though this James died the first of all the

the apostles, his brother John long survived them all. For he lived some time after his banishment to the isle of Patmos in the reign of Domitian, which was probably in A. D. 94. From the fate of James the rest of the apostles might see what they had to expect in this life; and yet it is evident that it did not operate as a discouragement to them. They all perished in the same persecuted cause, and most of them probably with no better treatment than he met with.

The other apostles of whom we have any writing left, viz. James and Jude, the former called *the brother of Jesus*, being either his natural brother of the same parents, or some near relation, breathes the same exalted spirit with Peter and John, earnestly exhorting his brethren to bear with patience and fortitude all the sufferings of this life, in the joyful expectation of receiving an abundant recompence in another.

"My brethren," he says, Ch. v. 7. "count it "all joy when ye fall into divers temptations, or ra- "ther trials," and again (i. 2.) "Blessed is the "man that endureth temptation, for when he is tri- "ed he shall receive the crown of life which the "Lord has promised to them that love him. (v. "12.) Be patient brethren unto the coming of the "Lord. Behold the husbandman waiteth for the "precious fruit of the earth, and has long patience "for it, until he receive the early and latter rains. "Be ye also patient, establish your hearts; for the "coming of the Lord draweth nigh," v. 7.

Jude,

Jude, to the same purpose, says, v. 21. "Keep "yourselves in the love of God, looking for the "coming of our Lord Jesus Christ unto eternal "glory;" and he concludes his short epistle in the following animating manner. "Now to him who is "able to keep you from falling, and to present you "faultless before the presence of his glory, with "exceeding joy, to the only wise God our saviour "be glory and majesty, dominion, and power, both "now and forever."

Except Matthew, the author of the Gospel which bears his name, no other of the twelve apostles were writers. They were not ambitious, nor indeed were those whose writings we have at all ambitious to be known to the world, and to be celebrated, as such. They only wrote what their circumstances, and those of their disciples, required; being content to wait for every honourable distinction till the return of their common master. We cannot, however, doubt but that their disciples, being, wherever they were, in the same circumstances with those to whom the epistles of the other apostles were addressed, they exhorted them on the same principles, referring them to that great day when the wicked will receive a due punishment, and the righteous an ample reward, and teaching them, as the other apostles did, not to place their affections on any thing in this world, or to be disturbed at any sufferings to which they should be exposed here; since they could only be for a time, and would bear no sensible proportion to the advantage they would derive from bearing them as became christians, that is with patience, fortitude, and with meekness, and without any ill will to their persecutors; and at the same time contributing every thing in their power to lessen the sufferings of their brethren.

How different is this disposition from that which is admired by the world at large, but how superior is it in the eye of reason, as it implies a greater command of temper, less governed by things present, and arising from a more extensive and enlarged view of things, the only proper evidence of our advance in intellectual above sensual life.

With this we, as well as all other animals, necessarily begin our career of existence, and the brutes never in general get much beyond it; but experience and observation lead *men* to extend their views, to reflect upon the past, and look forward to the future; and in this progress we pass from selfishness to benevolence, and from the contemplation of nature to the veneration and love of the great author of nature, both in doing and suffering, without any regard to what may be the consequence in this life, assured that by such sentiments, and such conduct, we shall not finally be any losers; but that when we shall have done the will of God, and have seen his goodness here below, an abundant entrance will in due time be administered to us in his everlasting kingdom and glory.

ON

THE CHANGE WHICH TOOK PLACE

IN THE CHARACTER OF THE APOSTLES

AFTER THE RESURRECTION OF

JESUS CHRIST.

[PART II.]

And when they saw the boldness of Peter and John, and perceived that they were unlearned and ignorant men, they marvelled, and they took knowledge of them that they had been with Jesus.

Acts. iv. 13.

IN the preceding discourse we considered the very remarkable change in the views and character of the twelve original apostles in general, and especially of those whose epistles furnish the proper evidence of it, viz. those of Peter, James, John, and Jude. We have seen that from being men of worldly ambition, expecting honours and rewards under the Messiah in this world, they suddenly abandoned every prospect of the kind, looking to nothing but a reward in heaven ; and that in the firm belief and expectation of *this*, they bore themselves, and exhorted others to bear, all the sufferings to which for the profession of christianity they could be exposed.

The

The clearness and energy with which they express themselves on this subject is most interesting and animating, and deserves as much attention in our days of *peace* as theirs of *persecution*. For if their situation required motives to patience and fortitude, ours requires constant admonition, lest the cares of this world should wholly exclude, as they naturally tend to do, all consideration of another. I shall, therefore, proceed to give as particular an account of the sentiments and exhortations of the apostle Paul on this subject as I did of those of the other apostles.

The change in the conduct, though not perhaps in the character, of Paul was as great, and as sudden, as that in the other apostles. Since from being a most violent persecutor of christianity, he not only became a christian himself, but a most active and successful propagator of christianity, especially in countries distant from Judea; and he seems to have gone through more hardships, and to have suffered more persecution of various kinds, on that account, during the course of a long life, than any other of the apostles; and at last, according to ecclesiastical history, he suffered martyrdom at Rome.

Of the worldly ambition of Paul we have no other evidence than the indirect one, which arises from his entering into the views of the leading men of his nation, and being the most active instrument they could employ; from which he would, no

doubt,

doubt, expect such rewards as men in power usually bestow; though at the same time his chief motive might be a genuine zeal for his religion, of the divine authority of which he entertained no doubt, and to which he thought the principles of christianity were hostile. He therefore believed it to be a duty which he owed to God and his religion, as well as to his earthly superiors, to do every thing in his power to suppress it. In other respects his general moral character was as unimpeachable as that of the other apostles. They were alike men of piety, integrity, and sobriety, though misled by the prejudices of their countrymen, who all expected a temporal prince in their Messiah, and therefore looked for such honours and emoluments as temporal princes have it in their power to bestow.

Thinking, as I have observed, that we in this age stand in as much need of admonition and exhortation concerning our interest in a future world as the primitive christians, I shall lay before you what the apostle Paul advanced on this subject, and we shall see it to be no less explicit and animating, and furnishing more information with respect to it than we find in the writings of the other apostles. In zeal and courage Paul yielded to no man, he derived his knowledge from the same source, viz. from Jesus in person, and his writings tend in an eminent degree to inspire the sentiments which he entertained himself. As the passages in the writings of the apostle relating to a future state are

numerous,

numerous, I shall recite them in the order of time in which they were written, beginning with the epistles to the Thessalonians, which were the first.

In Thessalonica Paul preached but a short time, probably not more than three weeks, Acts xvii. 1. &c. and so ill was he received there by the unbelieving Jews, who represented him and his companions as *men who turned the world upside down*, that he was persuaded to leave the place by night. The shortness of the time, therefore, would not admit of the converts there being fully instructed in all the principles of the new religion; and happily for us they had so far misunderstood what he had taught them concerning the resurrection, that he found it necessary to explain himself further on the subject, in an epistle which he wrote to them as soon as he reached Athens; since by this means we are acquainted with some circumstances concerning it which we could not learn from any other of the books of scripture.

It was a custom with the heathens to make loud lamentations over their dead, which, if they had any value for them while they lived, was natural, as they had no expectation of seeing them any more. This custom Paul thought unbecoming christians, and therefore he says (1 Thess. iv. 13.) "I would "not have you be ignorant, brethren, concerning "them that sleep, that we sorrow not as others who "have no hope. For if we believe that Jesus died "and rose again, even so they also who sleep in "Jesus

"Jesus will God bring with him. For this we say "unto you by the word of the Lord, that we who "are alive and remain unto the coming of the "Lord shall not prevent" (or rather shall not have any advantage over) "them that are asleep. For "the Lord himself shall descend from heaven with "a shout, with the voice of the archangel, and the "trump of God, and the dead in Christ shall rise "first. Then we who are alive and remain shall "be caught up together with them in the clouds, "to meet the Lord in the air, and so shall we ever "be with the Lord. Wherefore comfort one ano-"ther with these words."

This was, indeed, a source of consolation abundantly sufficient for the purpose, and peculiar to them as christians; so that they had no occasion to lament the death of their christian friends as the heathens did theirs, since they might depend upon seeing them again after the resurrection, and in circumstances far more advantageous than any they had known here.

It appearing that these christians at Thessalonica were still under some misapprehension about the doctrine of the resurrection, and especially about the *time* of it, conceiving it to be much nearer than it was, the apostle saw reason to address to them another epistle, not long after writing the first, and to correct the mistake they were under he says (2. Thess. ii. 1. &c.) "Now we beseech you, breth-"ren, by the coming of our Lord Jesus Christ,

"and

" and by our gathering together unto him, that ye " be not soon shaken in mind, nor be troubled, as " that the day of Christ is at hand. Let no man de-" ceive you by any means. For that day shall not " come unless there be a falling away first." He then proceeds to point out to them an antichristian power that was to arise in the church before the coming of Christ, from which they might gather that this great event could not be so near as they had imagined.

As this christian church at Thessalonica was soon exposed to much persecution, the apostle encourages the members of it to bear their sufferings with patience and fortitude, from the consideration of the abundant recompence that would be made to them at the coming of Christ, which would be as dreadful to their enemies, as it would be joyful to them.

" We are bound" he says, " to thank God al-" ways for you, brethren, as it is meet, because your " faith groweth exceedingly, and the charity of eve-" ry one of you all towards each other aboundeth ; " so that we ourselves glory in you in the churches " of God for your patience and faith in all your per-" secutions and tribulations that ye endure ; which " is a manifest token of the righteous judgment of " God, that ye may be accounted worthy of the " kingdom of God for which ye also suffer. Seeing " it is a righteous thing with God to recompence " tribulation to them that trouble you ; and to " you

"you who are troubled rest with us; when the
"Lord Jesus shall be revealed from heaven with
"his mighty angels, in flaming fire, taking ven-
"geance on them that know not God, and that o-
"bey not the gospel of our Lord Jesus Christ: who
"shall be punished with everlasting destruction
"from the presence of the Lord, and from the glo-
"ry of his power, when he shall come to be glori-
"fied in his saints, and to be admired in all them
"that believe."

With great reason did the apostle exhort these christians (1 Thess. iii. 3.) not to be moved by their affliction. "You yourselves," says he, "know "that we are appointed thereunto. For verily "when we were with you, we told you before, that "we should suffer tribulation, even as it came to "pass, and ye know."

If it was happy for us that the Thessalonian christians mistook the meaning of the apostle with respect to the resurrection, it is more so that those at Corinth perverted it by a false philosophy; because we derive more advantage from the conceit of the latter, than from the ignorance of the former, as it gave occasion to the apostle to explain himself still more fully on the subject in his epistle to them. For in this he leaves little that we could reasonably wish to know concerning it.

The christians at Corinth misled by the principles of the Greek philosophy, were disposed to treat the doctrine of a resurrection with contempt, as a most improba-

improbable thing, as it also appeared to the Gnostic christians, and imagined that the apostle in announcing it must have had some other than a literal meaning. They held *matter*, and the *body* which is composed of it, in great contempt, and thought it a happy circumstance for the immaterial soul to be delivered from it by death, so far were they from wishing for a reunion with it at the resurrection. But the apostle, who, with the Jews, expected no future life but in the supposition of a proper resurrection, paid no attention to this Grecian philosophy; and therefore he considered the disbelief of the resurrection to be the same thing with the disbelief of a future state altogether; saying (1 Cor. xv. 17.) "If Christ be not raised, your faith is vain, ye are "yet in your sins." For he justly observed that, if there be no general resurrection, there are no particular ones, not even that of Christ, whereas there was the most direct and abundant evidence of the reality of *his* resurrection, which is the assurance of ours.

On this account he particularly enumerates most of the appearances of Jesus after he was raised from the dead, and especially his appearing to more than five hundred of his disciples at one time, most of whom were then living, and could attest it. But the resurrection of Jesus is a pledge of ours. Consequently, the apostle calls him (v. 20.) *the first fruits of them that sleep*; the great *harvest*, to which he alludes by the mention of the *first fruits*, being the

the resurrection of all his followers. It has pleased God, he observes, that "as by man came death, so "by man also comes the resurrection of the dead, "and that as in Adam all die, so in Christ shall all "be made alive." And as all power is to be put into the hands of Christ, and all his enemies are to be subdued by him, the last of them is death.

After this he proceeds to answer several objections that were made to the doctrine of the resurrection, especially with respect to the kind of body with which men will rise; and he observes that as every kind of corn that men sow and reap is renewed after being buried in the ground, it will be the same with men, but with this advantage, that our future bodies will not be like the present ones, liable to corruption, disease and death; for that with respect to it they may be called spiritual, like the glorified body of Jesus.

The same advantageous change he observes will take place in those who shall be alive at the coming of Christ. "We shall not sleep, but we shall all "be changed in a moment, in the twinkling of an "eye, at the last trump. For the trumpet shall "sound, and the dead shall be raised incorruptible, "and we shall be changed. For this corruptible "must put on incorruption, and this mortal must "put on immortality." After this, in the language of triumph, he adds, referring to a passage in Isaiah, "O death where is thy sting, O grave where is thy "victory. Thanks be to God who gives us the "victory through our Lord Jesus Christ."

On this glorious doctrine he immediately grounds this natural exhortation. "Therefore my beloved "brethren, be ye steadfast, immoveable, always, a-"bounding in the work of the Lord, forasmuch as "ye know that your labour shall not be in vain in "the Lord." Indeed there cannot be any more powerful motive to the diligent practice of our duty, and a steady perseverance in it.

This was the great encouragement and support to Paul himself under all the trials that he underwent in the propagation of the gospel, as we see in this epistle. "If" he says, v. 32, "after the manner "of men I have fought with beasts at Ephesus, "what advantage have we if the dead rise not. Let "us eat and drink for to-morrow we die."

In his second epistle he has recourse to the same animating prospect as that which supported him under all his tribulations. 2 Cor. iv. 8. "We are "troubled on every side" he says, "but not distress-"ed. We are perplexed, but not in despair; per-"secuted, but not forsaken; cast down, but not "destroyed; always bearing about in the body the "dying of the Lord Jesus, that the life also of Jesus "might be manifested in our body. For we who live "are always delivered unto death for Jesus sake, "that the life also of Jesus might be made mani-"fets in our mortal flesh." ib. 16. "For this "cause we faint not; for though our outer man "perish, yet the inward man is renewed day by "day. For our light affliction, which is but for a "moment,

"moment, worketh for us a far more exceeding
"and eternal weight of glory; while we look not
"at the things which are seen, but at the things
"which are not seen. For the things which are
"seen are temporal; but the things which are not
"seen are eternal." "For we know that if our
"earthly house of this our tabernacle be dissolved,
"we have a building of God, a house not made
"with hands, eternal in the heavens." v. 1. &c.

There is a peculiar energy in all the epistles that Paul wrote from Rome, where he was two years a prisoner, expecting his condemnation or acquittal at the tribunal of the emperor, to whom he had appealed from his prejudiced judges in Judea. Then too he was far advanced in life, and sensible that his continuance in it could not be long. In these circumstances his epistles are like the dying advices of an affectionate parent, urging upon his children such considerations as he then felt would be of the most importance to them. And a view to a future state of rest and reward would naturally be uppermost in the mind of one who had laboured and suffered so much as he had done in the cause of christianity. Accordingly, we find that a view of this was constantly upon his mind, and that he was upon every occasion directing the views of his fellow christians to it.

In the christians at Philippi Paul had found his most generous friends, who, it appears, had been, more particularly attentive to him than those in other

ther places. The Philippians, as well as himself, had been exposed to peculiar hardships from their first reception of the gospel. But how light did he make of all his sufferings, thereby intimating that they ought to make as little account of theirs in the same cause, when he says, Phil. iii. 8. "Yea "doubtless and I count all things but loss for the "excellency of the knowledge of Christ Jesus my "Lord, for whom I have suffered the loss of all "things, and count them but dung that I may win "Christ, that I may know him, and the power of his "resurrection, and the fellowship of his sufferings, "being made conformable unto his death; if by "any means I may attain to the resurrection of the "dead" (Phil. iii. 8. &c.) "Our conversation is in "heaven, from whence also we look for a Saviour, "the Lord Jesus Christ, who shall change our vile "body, that it may be fashioned like unto his glo-"rified body, according to the working whereby he "is able even to subdue all things unto himself." Phil. iii. 20. &c.

We have no account of Paul ever preaching at Colosse, but by some means or other the gospel had been preached and received there, as indeed it soon was in all the cities of Asia Minor. To these christians the apostle now writes from Rome, and in his epistle he does not neglect to remind them of their great interest in a future state, as a recompence for all their good deeds and sufferings in this. "We give thanks to God and the Father of our "Lord Jesus Christ, praying always for you since

"we

"we heard of your faith in Christ Jesus, and the
"love which you have for all saints; for the hope
"that is laid up for you in heaven, whereof ye have
"heard before in the word of the truth of the gos-
"pel." Col. i. 3. &c.

Timothy was a favourite disciple and fellow labourer with this apostle, who, after travelling with him, as an assistant and an evangelist, resided at Ephesus, a city of the greatest note in Asia Minor, and the metropolis of Asia proper. This, therefore, was a station of peculiar importance; and accordingly the apostle, in the epistle which he wrote to him from Rome, which is the second (for the first epistle to him was written long before when Paul was at Corinth) takes great pains to encourage and animate him, urging more especially the consideration of their future glorious prospects. "Be not
"therefore," says he, (2 Tim. i. 8.) "ashamed of
"the testimony of the Lord, nor of me his prisoner.
"But be thou partaker of the afflictions of the gos-
"pel, according to the power of God, who has sav-
"ed us, and called us to a holy calling; not accord-
"ing to our works, but according to his purpose
"and grace, which was given us in Christ Jesus
"before the world began, but is now made manifest
"by the appearing of our Saviour Jesus Christ,
"who has abolished death, and brought life and im-
"mortality to light through the gospel."

As a farther encouragement to him, he expresses his own satisfaction in the near view of his death. "I suffer," he says, (ii. 9.) "as an evil doer, but

"the

"word of God is not bound," as he then was.
"Therefore I endure all things for the elect's sake,
"that they also may obtain salvation which is in
"Christ Jesus with eternal glory. This is a faith-
"ful saying, that if we be dead with him, we shall
"also live with him; if we suffer, we shall also
"reign with him. If we deny him, he also will
"deny us."

In this near view of death he rejoices in the prospect of it, as the termination of all those labours
which would entitle him to a glorious recompence
"For I am now ready to be offered, and the time of
"my departure is at hand. I have fought the good
"fight, I have finished my course, I have kept the
"faith. Henceforth there is laid up for me a crown
"of righteousness, which the Lord, the righteous
"judge, shall give me at that day; and not to me
"only, but to all them that love his appearing."
2 Tim. iv. 6.

From these weighty considerations he gives Timothy the most solemn charge to attend to his duty as an evangelist, with a view to this great reward.
"I charge thee before God (iv. 1.) and the Lord
"Jesus Christ, who shall judge the quick and the
"dead at his appearing and his kingdom. Preach
"the word, be instant in season and out of season,
"reprove, rebuke, exhort, with all long suffering
"and doctrine."

Titus was another disciple and fellow labourer
with Paul, and was by him stationed in the isle of
Crete. Here likewise he earnestly exhorts to diligence,

gence, reminding him, as he had done Timothy, of the hope of eternal life, which he says, (i. 2.) "God, "who cannot lie, has promised before the world be- "gan." The grace of God, has appeared unto all "men, teaching us that denying ungodliness and "worldly lusts, we should live soberly, righteously "and godly in this present world; looking for that "blessed hope, and the glorious appearing of the "great God and our Saviour Jesus Christ, who "gave himself for us that he might redeem us from "all iniquity, and purify to himself a peculiar peo- "ple, zealous of good works." ii. 11. &c.

The Jewish or Hebrew christians were from the first exposed to grievous persecution from their bigotted countrymen, and a great proportion of them appear to have been in low and distressed circumstances, so as to stand in need of the benefaction of the more wealthy Gentile converts. To these the apostle holds out the most comfortable prospects in futurity. "Here," he says, (Heb. xiii. 14.) "we "have no continuing city, but we seek one to "come." And again, (xii. 28.) "Wherefore, we "receiving a kingdom which cannot be moved, let "us have grace, whereby we may serve God ac- "ceptably, with reverence and godly fear."

We see in the language of the apostles, and in their sentiments and conduct, which corresponded with it, the infinite advantage that christians, and even unlearned christians had over the most enlightened of the heathens, with respect to the troubles of life and the fear of death, in consequence of the firm be-

lief

lief of the former in the great doctrine of a future state, which was not only to be the termination of all their sufferings, but, under the righteous moral government of God, a certain means of obtaining an abundant recompence for all their sufferings in the cause of virtue here, whereas the heathens had little knowledge of any moral government of God, or of a providence here, and no knowledge at all that could be of any practical use of a future state. To them all beyond the grave was absolute darkness, but to christians it is the most resplendent light.

The christian sees the hand of God, of his God and father, in every thing that befalls him here; and he expects a greater display of his perfections, and more evident and uninterrupted marks of his favour hereafter. These views enable him to consider all the troubles of life as a part of that excellent and benevolent discipline which is to prepare him for future happiness, a discipline which he is taught to believe it as necessary to him, as the controul and discipline of a child is to his acquiring the proper sentiments and conduct of a man; qualifying him to be happy in himself, and disposed to make others so; which without this controul and discipline in the time of childhood and youth, it was impossible that he should be. And the near approach of death, which at the best cannot but afford a gloomy prospect to a heathen and an unbeliever, is consequently regarded by him not as an object of alarm, or despondence, but a source of joy and triumph; so that when he leaves the world, which he believes to

be

be at the call and appointment of him that made him, and sent him into it, he can with the apostle sing the triumphant song, O death where is thy sting, "O grave where is thy victory." Thanks be to God who gives us the victory, through our Lord Jesus Christ.

The difference between the moral writings of the heathens, and those of the apostles, to the advantage of the latter, cannot but appear upon the slightest attention. As these, besides being superior in point of clearness, have, from the fulness of their persuasion on the subject, which the heathens had not, infinitely more of animation; so that the perusal of their writings cannot fail to excite the same sentiments in others.

As I have purposely confined myself to the subject of courage and perseverance, in bearing sufferings of every kind, and even persecution unto death, from the prospect of a future glorious reward which was wholly unknown to the heathens, I shall now recite a few passages from the epistles of Paul, in which mention is made of the sufferings to which he was exposed, and of his magnanimity in bearing them, without any immediate view to a future reward, though no doubt it was constantly on his mind.

At Corinth the christians seem to have been so numerous, and respectable, in the time of the apostle, or their fellow citizens so much more civilized than those of many other places, that they were less exposed to persecution than the christians in other places; and they had among them some eloquent declaimers, who seem to have derived pecuniary emolument from their harangues. The apostle, therefore, represents their situation as enviable with respect to that of other churches, and on this account he seems to have chosen to describe his own situation by way of contrast with theirs. "Now," says he, (1 Cor. iv. 8.) ye are "rich. Ye have reigned as kings without us, and "I would to God that ye did reign, that we also "might reign with you." And he immediately adds the following affecting account of his own situation.

"I think that God has set forth us the apostles "last, as it were appointed to death; for we are "made a spectacle to the world, and to angels, "and to men.—Even to this hour we both hun-"ger and thirst, and are naked, and are buffetted, "and have no certain dwelling place, and labour, "working with our own hands. Being reviled, "we bless; being persecuted, we suffer it; being "defamed, we intreat. We are made as the filth "of the earth and are the offscouring of all things "unto this day." 1 Cor. iv. 9. &c.

This was in his first epistle to this church. In the second, which was written not long after it, he still reminds them of his sufferings, to which it is probable they had not been sufficiently attentive. "We would not, brethren, have you ignorant of "our trouble which came to us in Asia, that we "were pressed out of measure, above strength, "so "so that we despaired even of life. But we had "the sentence of death in ourselves, that we should "not trust in ourselves, but in God who raises the "dead; who delivered us from so great a death, "and doth deliver, in whom we trust that he will "yet deliver us." 2 Cor. i. 8. &c.

"In all things approving ourselves the ministers "of God, in much patience, in afflictions, in neces-"sities, in distresses, in stripes, in imprisonments, "in tumults, in labours, in watchings, in fastings; "by honour and dishonour, by evil report and "good report: as deceivers, and yet true; as un-"known and yet well known; as dying and behold "we live; as chastened and not killed; as sorrow-"ful yet always rejoicing; as poor yet making ma-"ny rich; as having nothing and yet possessing all "things." 2 Cor. vi. 4. &c. As a contrast of his situation with that of the eloquent speakers in this church of Corinth, who seem to have been much at their ease, he gives the following affecting account of his labours and sufferings. 2 Cor. xi. 23. &c. "Are they ministers of Christ, I am more. "In labours more abundant, in stripes above mea-"sure, in prisons more frequent, in deaths often. "Of the Jews five times received I forty stripes "save one. Thrice was I beaten with rods, once "was I stoned, thrice I suffered shipwreck, a night "and a day I have been in the deep. In journey-"ings often, in perils of water, in perils of robbers, "in perils by my own countrymen, in perils in the "wilderness, in perils in the sea, in perils among

"false

" false brethren. In weariness and painfulness, in
" watchings often, in hunger and thirst, in fastings
" often, in cold and nakedness. Besides those
" things that are without, that which cometh upon
" me daily, the care of all the churches. Who is
" weak and I am not weak? who is offended and I
" burn not? If I must needs glory, I will glory of
" the things which concern my infirmities. The
" God and Father of our Lord Jesus Christ, who is
" blessed for evermore knoweth that I lie not. In
" Damascus the governor under Aretas the king
" kept the city of the Damascenes with a garrison,
" desirous to apprehend me; and through a win-
" dow, in a basket, I was let down by the wall, and
" escaped his hands."

In his epistle from Rome, written in the near prospect of death, after enduring, as we have seen, such a series of hardships as few men have ever gone through, he thought proper to remind the churches to which he wrote of what he had suffered, that they might not be surprised, or discouraged, if they met with no better treatment in this world than he had met with.

To the Ephesians he says, Ch. iii. 13. " Where
" fore I desire that ye faint not at my tribulation
" for you, which is your glory;" intimating that so far from being discouraged, or ashamed, they ought to be proud of these proofs of his affection for them, and of his zeal in the common cause.

To the Colossians he says, to the same purpose, Ch. i. 24. " I rejoice in my sufferings for you,
" and

"and fill up that which is behind of the afflictions "of Christ in my flesh, for his body's sake, which "is the church." As if a certain portion of suffering had been necessary to establish christianity and as if that of Christ had not been sufficient, he took the remainder upon himself. The same idea occurs, though not so distinctly, in his epistle to the Galatians, written long before this. Gal. ii. 20. "I am crucified with Christ; nevertheless I live; "yet not I, but Christ liveth in me. Wherefore "let no man trouble me, for I bear in my body the "marks of the Lord Jesus." vi. 17.

The christians at Philippi had suffered much. Writing to them from Rome, he expresses the greatest indifference and contempt of all that could befal him. "In nothing," he says, (Ch. i. 20.) "shall I be ashamed, but that with all boldness, "as always, so now also, Christ shall be magnified "in my body, whether it be by life or by death. "I have learned," he says, (iv. 11.) "in whatever "state I am, therewith to be content. I know both "how to be abased, and how to abound. Every "where, and in all things, I am instructed both to "be full and to be hungry, both to abound and to "suffer need. I can do all things through Christ "who strengthens me."

The general sentiment of the duty of patience and fortitude under the evils of life may, no doubt, be found in the writings of Marcus Antoninus, Seneca, and other heathens; but the feelings they convey are very different, quite feeble and inefficacious.

cious. The heathens could not have the same *motives* to patience and fortitude. Those of christians are infinitely more efficacious, and far more *natural*, as they are taught to look beyond them to objects which in similar cases do not fail to enable men to bear hardships of any kinds, viz. to a certain advantage accruing from them, and to which they are necessary. If the christian suffers here, especially in the cause of virtue and truth, he is taught to expect a certain recompence in a future state. Compared with this, the patience and fortitude of heathens, especially in the near view of death, cannot be much more than mere obstinacy, arising from the consideration of the necessity of bearing what they cannot avoid; and therefore of the folly of complaining where it cannot answer any good end.

Let the writings of the Stoics on this subject be compared with those of the apostles, and the difference must be striking. In the sufferings of christians we see there is a source of joy. Paul speaks of *rejoicing in tribulation*, but for this the Stoic could not have any motive. The apostles did not deny that painful sufferings were evils. They acknowledge that they were not in themselves *joyous but grievous*, but they *worked out for them a far more exceeding, even an eternal weight of glory*. According to the apostles, it is only *for a time*, and *if need be, that we are to be in sorrow through divers trials*, and to the end of this time they were well able to look, and, like their master, for

for *the joy that was set before them*, they endured every affliction, and even the pains of death itself.

Let us now hear Marcus Antoninus on the subject of the fear of death, to which he frequently adverts in his *Meditations*, and from which we may infer that it was much upon his mind. After enumerating the duties of life, which he says, "every "man is under obligation to discharge," he says, (ii. 17.) "he must expect death with a benevo-"lent and calm mind, as a dissolution of those "elements of which every animal consists. And "if nothing uncommon happen to these elements, "and they be only changed as all elements conti-"nually are, into others, why should we dread "the event, or be disturbed at that change and dis-"solution which is the lot of all. For it is ac-"cording to nature, and nothing that is natural is "an evil."

How poor is the consolation which this language holds out compared with that of the apostles, which has now been recited. His reasoning about the indifference with which we should regard the duration of life is unsatisfactory; and indeed manifestly absurd, if life be of any value. "If any "of the gods," he says, (iv. 47.) "should tell "you that you should die either to-morrow, or "the day following, you would not be disturbed "at it; unless you were of a very cowardly and "abject disposition. The difference between to-"morrow and the day following is indeed a trifle; "but for the same reason you should not make any

"account

" account of the difference if it should be either to-
" morrow, or a thousand years hence." I doubt not, however, but that if the emperor himself had the choice of dying either after one more day of life, or of living, I do not say, a thousand years, but to the usual time of human life, he would not have hesitated to show, by his actual choice of the latter, that he thought it was not a matter of so much indifference as in his writings he represents it.

How thankful, then, should we be for the gospel, which gives us such an unspeakable advantage over the most enlightened of the heathens with respect to what must interest all men the most, the troubles of life, and the fear of death. Under these the heathens could at the best only *acquiesce*, as in things that were unavoidable; and being, as Antoninus says, *agreeable to nature*, must be the best with respect to the whole system; but not for them in particular. They had nothing to look to beyond the business and the troubles of this life, and no hope at all after death. And their arguments for patiently acquiescing under the evils of life, and in the view of death, would never have any weight with the bulk of mankind, and whatever they might pretend, could only be affected by the philosophers themselves. Whatever they might teach, or write, they must have *felt* like other men in the same circumstances, having no more expectation of surviving death, or ever seeing any better state of things, than other men.

Being

Being then through the goodness of God possessed of this superior knowledge, this *treasure* so long hidden from the greatest part of the world, this *pearl of great price*, let us value it in proportion to its real worth, converting this *knowledge* into useful *feelings* and *practice*, by living agreeably to the light with which we are favoured. Otherwise, it would have been better for us to have continued ignorant heathens, as we should then have had less to answer for; and woe will be to those who when this *light is come into the world* shew by their conduct that *they love darkness better than light because their deeds are evil.* To our christian knowledge, let us, with the apostle, add all the proper virtues of the christian life. These exceeding great and precious promises are given to us, that, as the apostle Peter says, *we may thereby become partakers of a divine nature*, having escaped the corruptions that are in the world. *Giving all diligence*, as he exhorts, *let us add to our faith virtue, and to virtue knowledge, to knowledge temperance, to temperance patience, to patience godliness, to godliness brotherly kindness, and to brotherly kindness universal charity. If these things* as he says, *be in us and abound, we shall not be unfruitful in the knowledge of our Lord Jesus Christ.*

Let us then, my christian brethren, *give diligence to make our calling and election sure, for*, as the same apostle adds, *if we do these things we shall never fail; for so on entrance will be administered unto us abundantly into the everlasting kingdom of our Lord Jesus Christ.*

ERRATA

IN THE SECOND VOLUME.

PAGE	LINE		
360	6 b	*for* L'uesceque	*read* Q'uestceque
403	7	.... Scyrius	...... Syrius
422	2	.... Confervu	...... Conferva
.....	3 b	.... Hydortids	...... Hydatids
423	3	.... Nydra	...... Hydra
426	5	.... Terms	...... Terrors
426	12	.... Acccrate	...... Accurate
444	5	.... Flower	...... Slower
674	2 b	.... Alexandrius	.... Alexandrinus
736	1 b	.... Bavorius	...... Barônius

Printed in the United States
By Bookmasters